brilliant

copywriting

brilliant

copywriting

How to craft the most interesting
and effective copy imaginable

Roger Horberry

PEARSON
Prentice Hall

Harlow, England • London • New York • Boston • San Francisco • Toronto • Sydney • Singapore • Hong Kong
Tokyo • Seoul • Taipei • New Delhi • Cape Town • Madrid • Mexico City • Amsterdam • Munich • Paris • Milan

PEARSON EDUCATION LIMITED

Edinburgh Gate
Harlow CM20 2JE
Tel: +44 (0)1279 623623
Fax: +44 (0)1279 431059
Website: www.pearsoned.co.uk

First published in Great Britain in 2009

© Pearson Education Limited 2009

The right of Roger Horberry to be identified as author of this work has been asserted by him in accordance with the Copyright, Designs and Patents Act 1988.

ISBN: 978-0-273-72734-7

British Library Cataloguing-in-Publication Data
A catalogue record for this book is available from the British Library

Library of Congress Cataloging-in-Publication Data
A catalog record for this book is available from the Library of Congress

The publishers would like to thank MovieMaker Magazine for permission to reproduce the extract on page 58.

10 9 8 7 6 5 4 3 2 1
13 12 11 10 09

Illustrations by Mark Gravil, sc.sixty5@dsl.pipex.com
Typeset in 10/14pt Plantin by 3
Printed and bound by Henry Ling Ltd, at the Dorset Press, Dorchester, Dorset

The publisher's policy is to use paper manufactured from sustainable forests.

For Lindsay, Lucy, Lotte and George.

Now you know what I do all day.

Contents

About the author

I'm Roger Horberry, I'm married with two children and I live in North Yorkshire. By day I'm a freelance copywriter working for various design, branding and advertising agencies. By night I make wilfully obscure music. Over the last 25 years I've released ten CDs that together have earned me literally hundreds of pounds. *Brilliant Copywriting* is my first full-length book.

Write to me at roger@rogerhorberry.com.

Acknowledgements

I'd like to express my considerable gratitude to John Simmons for his endless encouragement and support, my wife Lucy Wilson for putting up with me being weird for weeks at a time, the good people at www.26.org.uk (a not-for-profit writers' group dedicated to inspiring a greater love of words in business and in life), and my marvellous copywriter interviewees in Part 3. A tip o' the hat to you all.

How to use this book

Read this bit first – it'll help you get the best from what follows.

To get a feel for the job of copywriting and to immerse yourself in the philosophy of being a brilliant copywriter, look at Part 1. It concentrates on the theory of brilliant copywriting, although there are plenty of practical tips scattered throughout.

If you want practical tools to improve your writing, go straight to Part 2. It describes hands-on techniques for brilliant copywriting. It ends with a series of worked examples that show what brilliant copywriting means in practice.

Finally, for an insight into how top copywriters work, go to Part 3. It contains a number of in-depth interviews with successful practitioners as well as summaries of their main ideas.

Call me an old stick-in-the-mud, but starting at the beginning and reading to the end is probably the best way to digest *Brilliant Copywriting*. However, this book is designed to be read in different ways by different readers. Anyone new to the world of copywriting will get a good overview from Part 1. If you're an established copywriter and you'd like to top up your toolbox of techniques then go to Part 2. Experienced writers could start with the interviews in Part 3 before heading back into the body of the book. Basically, how you use it up to you.

Preface

In the beginning

In the late 1890s a Canadian Mountie named John E. Kennedy hung up his hat, handed in his gun and did what any rock-jawed law enforcement operative would do in similar circumstances – he got into advertising. Not just any advertising, mind. Kennedy was concerned with copy. The advertising agencies of his day employed plenty of people called copywriters, but their craft was far from fully defined. Kennedy had thought long and hard about how words could be made to work harder. He realised the burgeoning advertising industry was missing a trick and that he was the man to put things right. So that's exactly what he set out to do.

Fast-forward a few years to a May evening in 1904. Kennedy is seated in a Chicago saloon. He scribbles a brief note to the directors of an advertising agency that occupied the upper floors of the building. According to legend the note read:

I am in the saloon downstairs. I can tell you what advertising is. I know you don't know. It will mean much to me to have you know what it is and it will mean much to you. If you wish to know what advertising is, send the word 'yes' down by the bell boy.

Somehow the note made it to Albert Lasker, one of the agency's junior partners. Lasker was getting ready to go home for the evening but was sufficiently intrigued by its chutzpah to invite the impudent Kennedy upstairs for a chat. Most people would

**John E. Kennedy ponders his
theory of copywriting.**

have written Kennedy off as an over-refreshed punter and asked security to have a quiet word. Not Lasker. To his credit (and indeed immense enrichment) he saw something in Kennedy's scrawl worthy of investigation, and by the time the two of them parted in the early hours of the next day nothing in copywriting would be the same again.

During that meeting Kennedy revealed to Lasker his big insight – that copywriting is 'salesmanship in print'. Nothing more, nothing less. He complained that adverts didn't need to be 'charming or amusing or necessarily pleasing to the eye' but instead they should be a 'rational, unadorned instrument of selling'. The copy should talk neither down nor up to its reader; instead it should address them in a way that left them 'open to appeals made by sensible arguments'.

All of this sounds eminently sensible until you start to dig a little deeper. Kennedy's hard-headed 'salesmanship in print' is a great start, but it's only half the story. Take this approach too far and pretty soon copywriting becomes cold and unconvincing. People rarely buy for wholly rational reasons, and in many cases copywriting that is 'charming or amusing' *can* outperform its rational relation. Not that you'd know it from the lacklustre copy put before today's long-suffering public. The sad fact is that although selling and salesmanship derive much of their power from brilliant copywriting, most copywriting is far from brilliant. In fact most of it doesn't really work. Why? Because no one reads it. Why? Because it's boring. Or bad. Or just plain irrelevant.

> rational is good;
> emotional is better

Sorry, but it's true.

Most copy isn't written with the reader in mind. It doesn't answer their questions, speak their language or tickle their fancy. And copy that isn't written for its reader is almost certainly destined to fail.

Trying to browbeat your audience into submission through a toxic mix of dull writing and endless repetition is a desperate, wasteful strategy. The answer is to *Make It Interesting*, my highly presumptuous update to Kennedy's three-word formula for success. If this book has one main message it's that brilliant copywriting almost always appeals to its readers for reasons other than pure reason – in other words Kennedy's rationalist 'reasons to believe' approach must be tempered with whatever it takes to capture the reader's imagination, and that usually means introducing something softer, more emotional, and above all *interesting*.

> brilliant copywriting almost always appeals to its readers for reasons other than pure reason

About this book

Who it's for

In the past copywriters worked for advertising agencies; today they crop up in all manner of creative companies. Consequently this book is aimed at *anyone* who uses words to persuade or sell. You might work in advertising, design, branding, PR, journalism, marketing or sales; equally you might be a manager interested in how copy works, a student who wants to break into copywriting, or a teacher looking for course material. All are welcome here. The only thing that unites the many different tribes for whom I wrote this book is that they want to improve how they persuade in print.

That last point is important. The copywriting I describe is primarily for page and screen; if you need to know about the specialised task of copywriting for TV, radio, outdoor and so on there are several excellent books listed in the Bibliography and Further Reading section at the back. This book focuses on what I'll describe as marketing copywriting. However, I happen to believe that writers working in design, branding, business and so on have much to learn from our colleagues in advertising, a hunch that I hope is borne out in the various interviews with advertising people towards the end. No group has a monopoly on knowledge; the more we share, the more we all benefit.

What's inside

This book is divided into three parts: Background, Method and Interviews. Where possible the emphasis is on *doing*. That's one of the great things about copywriting – you *create* stuff. In my opinion the pleasure of turning in a great piece of copy – anything from a short phrase to a lengthy think-piece – can make all the dead ends, rejections, delays and general nonsense that come with the job worthwhile.

Inevitably this book contains all sorts of, well, not *rules*, let's call them principles and methods. Some free spirits will say such things act as a brake on brilliant copy. That's a nice idea, but like many such ideas it's bunk. Without a sound knowledge of the nuts and bolts of brilliant copywriting the chances are that a writer, even a naturally talented one, will produce a load of self-indulgent poppycock. It's the old saw about having to know the rules in order to break them. So if at times I sound slightly didactic I make no apology.

On being brilliant

For me, 'brilliant copy' is simply copy that *works* – it grabs the reader's attention, gets its message across and persuades them to act – and it does all that with economy, wit and style. At the end of Part 2 I've included examples of just such copy, drawn from work I've done over the years. This is not – repeat, *not* – because I'm incredibly pleased with myself. One peek into the seething pit of torment that is my creative soul would soon convince you otherwise. No, it's simply that this material helps me make the points I need to make without any wearisome copyright issues.

> brilliant copywriting grabs the reader's attention, gets its message across and persuades them to act

And while we're on the subject of confessions, let's have another. Can this book – any book – *really* make you a brilliant copy-writer? I truly don't know. What I do know is that what follows will give your copywriting career a rocket-assisted launch. You'll learn how some really great writers go about their business and you'll pick up all sorts of tips and tricks that would otherwise take years to acquire. The rest – realistically – is up to you. That's fair enough, isn't it?

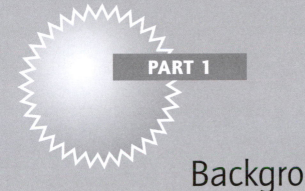

PART 1

Background

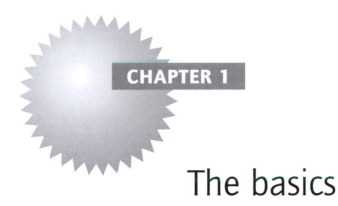

CHAPTER 1

The basics

What exactly *is* copywriting?

It's an obvious question, but like many obvious questions the answer isn't exactly straightforward. One copywriter might be working on a script for a corporate presentation, the next perfecting pack copy for posh sausages, while a third is crafting a TV campaign for a charity. So one legitimate answer to the question 'What exactly is copywriting?' is 'Marvellously varied'.

Another, slightly more obliging, answer to the above question is that copywriting is the job of using the right words, to say the right thing, to the right people, to get the right response. Those 'right words' can appear in any number of places, including adverts, annual reports, articles, brochures, case studies, company and product names, datasheets, direct mailers, flyers, leaflets, letters, newsletters, packaging, posters, presentations, straplines, websites and plenty more besides.

> copywriting is the job of using the right words, to say the right thing, to the right people, to get the right response

Another answer – touched on in the anecdote about John E. Kennedy in the Preface – is that copywriting is the business of selling with content. I use 'sell' in the loosest sense because copywriting is above all about *persuasion*. Copywriters aim to convince their readers of the merits of a particular product, service, argument or whatever, and then get them to act accordingly. Sometimes that means an overt, 'buy me'-type sell; at other times it's about getting them to buy into your way of thinking and see things from a particular point of view. This emphasis on persuasion doesn't mean copywriting is intrinsically less satisfying than other forms of writing. On the contrary, focusing your thoughts into an intense beam that ignites your reader's imagination and causes them to act in some way is every bit as challenging as writing fiction, and the chances are you'll be in work a lot more and paid rather better.

Put all this together and for me a copywriter is a professional persuader responsible for creating a specific message for a specific audience for a specific purpose. Copywriters need imagination to create interesting ideas, and craft skill to capture those ideas in a form of words that appeals to their readers. It's the copywriter's job to answer the unspoken 'So what?' in their readers' minds. If you can make your case in a way that gets through your audience's mental defences then you've a chance of making the sale I talked about earlier.

Some copywriters hold staff positions while others are freelance, a way of working that many copywriters try at some point in their career. It seems to suit the misanthropic streak in our collective character and often offers more variety (and indeed more cold, hard cash) than a staff position. On the other hand, freelancers tend to take whatever work comes their way regardless of its quality, and job security is the stuff of dreams (and occasionally nightmares).

Despite claims of rampant individualism many copywriters exhibit common character traits. These include obsessive curiosity about all manner of odd subjects, a magpie-like tendency to steal shiny words and phrases for use another day, a sense of humour, the ability to buckle down and work hard when the occasion demands, and the happy knack of thinking visually.

This last one is especially important within the context of advertising and design – a quick flip through any D&AD Annual reveals that many of the most effective and successful entries feature very little writing, yet they communicate with the utmost efficacy. It's a cliché to say that 'sometimes the best copy is no copy at all' but that doesn't make it any less true. The copywriter who can think visually will always be able to get the message across, regardless of their particular medium.

In a nutshell:

- Copywriting is about selling with content.
- It's about using the right words, to say the right thing, to the right people, to get the right response.
- A copywriter is a professional persuader responsible for creating a specific message for a specific audience for a specific purpose.

What copywriters do all day

Clearly there's a certain amount of staring into space to be done. Likewise tea-making and chatting. But in the end most copywriters spend most of their day behind a computer either writing, researching or thinking.

If you work for an agency you'll also attend meetings (both internal and external), some of which will actually be useful. If you're a freelancer you'll attend fewer meetings but more speculative interviews where you show your portfolio or book to those transcendent individuals (typically creative directors) who commission freelance copywriting. But mainly the copywriter's day is composed of reading, writing and thinking. And staring into space.

Now for a few more observations on the copywriter's lot.

Assimilation is everything

Improbable as it may sound, space gazing (as my brother-in-law calls it) is an essential part of the copywriter's day. Giving yourself sufficient thinking time is an important part of creating brilliant copy. Ideally you should aim for a 1:1 ratio between intake (typically reading, watching or listening) and assimilation (typically thinking, understanding and connecting), although that's often hard to achieve. I'll talk about the process of having ideas later in this book, but for now I'll leave it by saying that information itself is pretty useless, it's *understanding* – the gilded progeny of thinking and assimilation – that makes knowledge useful. Scrimp on this and your chances of writing brilliant copy are sorely reduced.

> giving yourself sufficient thinking time is an important part of creating brilliant copy

Getting the thinking right

The process of identifying, improving and capturing ideas is the dynamic behind much copywriting, and it's here that copywriters make can make a real contribution to the creative process. The act of getting something down on paper or screen forces the copywriter to explore the limits of the idea, to test its integrity and to fix any problems encountered. I can't overemphasise how important this idea of 'getting the thinking right' can be – if a piece of copy isn't conceptually robust then its persuasive mission is unlikely to succeed. Get the foundations right and whatever follows has at least a fighting chance of fulfilling the brief.

First in, last out

Even the most brilliant copywriter won't make much impact if they're only brought in at the last minute. Asking some poor copy monkey to convert blocks of *lorum ipsem* embedded in finished designs into real words isn't helping anyone – the result will inevitably be unsatisfying and superficial. No, in order to do their job well a copywriter needs to get involved at the start of the creative process and stay involved until the end.

Here's how it works. A copywriting project typically begins with one or more client meetings. If there's even a hint of chemistry between client and creative there'll be plenty of ideas flying about. At this early stage everything matters, even the most gruesome clichés (of which there will be plenty), because the smallest remark or aside may say something fundamental about how the client views their product, their market, their customers, their employees and themselves. Someone needs to pan for gold in this stream of consciousness. Someone needs to spot the good stuff, develop it and present it back to everyone involved for further refinement. In short, someone needs to *turn the ideas into words*, and naturally that someone is the copywriter. It makes sense – words are quick, easy and cheap to work with, ideal for what industrial designers and software engineers call rapid prototyping. Only when everyone involved is vaguely happy with the ideas being discussed is it appropriate to unleash other creative personnel, typically designers or art directors.

Once that's happened copywriters can provide a link between the thinking and the doing phases of a project. Writing provides continuity and ensures coherence in the final execution. It helps matters enormously if a project's copywriter is the same person who contributed at the thinking stage, not least because they'll then be able to sell the final idea to the client in a way that both assures and excites. As I say, a copywriter should be first in and last out.

Making it through meetings

Finally, a tip to get you through your day (and indeed your career) relatively unscathed – be yourself as much as possible. In particular don't worry about the impression you're creating in meetings – the chances are everyone else in the room will be worrying about exactly the same thing. And while we're on the subject of meetings, I strongly advise you to speak up as a way of signalling that you're an active part of what's going on. Don't be the strange, silent person with no apparent function – if a thought occurs to you and it's not wholly imbecilic or obscene, my advice is to blurt it out – you'll feel better and you might well be spot on. Get this right enough of the time and you'll become known as that helpful person with lots of ideas. And as we'll see later on, having a steady stream of good ideas is a big part of what brilliant copywriting is all about.

In a nutshell:

- Copywriters turn ideas into words.
- Get the thinking right and then capture it in an effective form of words.
- Information is useless; it's understanding that makes knowledge useful.

Dealing with dullness

This book is about the reality of copywriting, so it is my distressing duty to tell you that some of the writing you'll be asked to do – certainly in the early years of your career – will be downright dull. Even established copywriters do more bread and butter work than they'd like. Luckily there are a couple of things you can do to mitigate the corrosive effects of this situation.

First it's usually possible to find an interesting angle in even the blandest brief. The trick is to believe – really believe – that there's a cool solution just waiting to be found. Failing that, perhaps you can use this job to try out a new research technique, writing schedule or software package. Perhaps you can work in a different way, in a different location, or with different people. Perhaps it doesn't need any copy. Perhaps it should be nothing but text. Perhaps the title is everything. Perhaps it doesn't need one. You get the idea. It's up to you to find the seam of gold, no matter how narrow, and mine it for all you're worth.

Accept what you can't change

That's fine if the client is up for it, but what do you do when they specifically ask for dull work? They won't put it like that but their intention will be clear. One supermarket I worked for rejected some posters I'd done as 'too good for us', and no, they weren't being sarcastic. The answer is to acknowledge these limitations as beyond your control and work within them, doing the best job you can by saying what's good about the product in the clearest, most honest, most informative way possible. In

> do the obvious thing extraordinarily well

these situations it's a question of doing the obvious thing extraordinarily well. Do it cleanly, effectively and authentically and your readers will buy your passion (if not your purple prose). In short I'm preaching a kind of enlightened realism that acknowledges not every job comes with a potential award attached.

Faced with a choice, do both

Finally, if something feels wrong with a job or brief then the chances are it is. In this situation one option is to do your best with the brief as it stands, *and* do whatever you think is right as an alternative. Explain the situation to your boss and get him or her on your side. Don't phrase it along the lines of 'I just didn't fancy doing it their way'; explain *why* the work you're being asked to do will be a less effective piece of communication that your proposed solution. The client may reject your version out of hand and go with the tedious nonsense they asked for, but your self-respect will be intact and the idea you came up with might live to fight another day (you do save all your rejected ideas for recycling on future jobs, don't you?).

In a nutshell:

- Find your own interest – make it work for you.
- In some situations all you can do is say what's good about a product in the clearest, most honest, most informative way possible.
- Do the obvious thing extraordinarily well.
- Make your work as good as it can be while acknowledging that not every job has an award attached.
- Try doing a more satisfying alternative for your own benefit.
- Keep all your rejected ideas.

Dealing with clients

If there are no bad jobs, there are – just occasionally – bad clients. Well, not *bad* as such, more indecisive, timid, irrational and unreliable. Few, if any, behave this way in a deliberate attempt to ruin your day. They'll almost certainly have reasons for acting the way they do, reasons that are probably far beyond their control. In other words they can't help it. But we can help them, and that's what this section is about.

Use the brief to beat them off

Let's say a client is rejecting idea after idea that you've lovingly crafted and put before them. You listen to their comments, you revise the work accordingly, but still they won't say yes. Meanwhile time is slipping away, budgets are disappearing and tempers are fraying. Clearly you need to find out why they're unhappy with your work, and the way to do that is to *use the brief*. Assuming you've read and interpreted it correctly, you can request a meeting with the client to point out – with the greatest respect – that *what you're doing is what we agreed*. Be courteous but firm and make sure you've plenty of facts to support your case. This may result in changes to the brief, but that might be all you need to achieve a breakthrough. At the least it'll show the client that you're serious about solving their problem.

Invoking the brief is also the way around the hateful 'I'll know it when I see it' predicament. This is when a client rejects your work with some gibberish along the lines of, 'Come to think of it, I'm not quite sure what we need here, but I'll know it when I see it.' This approach has all the efficiency of wandering around The British Library scanning the shelves for a particular book rather than going to the index. It's a whirling vortex into which time, money and morale are sucked. If a client responds in this way it's tantamount to saying the brief is irrelevant. They might be right of course, so use this occasion to get to the bottom of what they really want. Whatever you do, don't allow yourself to be messed around as a result of client indecision. Stick to the brief, or insist on a new one, or get out. That's easier for me to write than for you to do (especially if you're a staff writer), but unless you're resolute you risk becoming their drudge. And drudges don't do brilliant copywriting.

Ask what's wrong, not what's right

If you're making progress it sometimes pays to deliberately ask for criticism from the client. If they play along you'll learn more about your craft and they'll get an even better piece of work. You don't have to do your efforts down, just ask 'Is there anything I could do to make this better?' or words to that effect. If you ask something like 'What do you think?' the chances are they'll reply in the affirmative even if they have reservations. Asking what more you can do shows how willing you are to do that extra bit to make a piece of work brilliant. It's a 'good is the enemy of great' thing.

> ask, 'Is there anything I could do to make this better?'

Find out what they really want

One of the best ways to make clients happy is to find out what they want – I mean *really* want. You might think that's what the brief is for, but in reality the chances are you'll have to read between the lines. In any event, find a form of words that capture what the client wants to feel about their company but can't quite articulate. If you can clarify their thoughts like this then they'll love you for it. I helped rebrand a London housing association whose implicit mission was to improve people's lives, and came up with the strapline 'Love where you live'. That simple little phrase gave focus to all their efforts and reflected exactly why they did what they did. I've still got the tee shirt.

Take the rap

Things occasionally go horribly wrong. If that happens and you're involved, I recommend you 'fess up. That doesn't mean grovelling, it means showing an honest capacity for self-criticism and an understanding of the responsibility that being a brilliant copywriter entails. Explain what happened and how you'll put it right. A useful trick here is to use the future tense as a way of defusing criticism. Say something like, 'Obviously it's unfortunate that the copy I wrote led to your lengthy imprisonment and the collapse of your company, but what's *really* important here is sorting out how we can prevent this from happening again.' It shifts the focus forward towards a sunny future. Does this rhetorical trick contradict my advice to accept responsibility for mistakes? Maybe, but there's no sense in needlessly offering yourself up as a whipping boy (unless you're into that sort of thing).

How much work should you show?

Sometimes your problem is an embarrassment of riches. If you've been particularly prolific then you're faced with the problem of what to show the client. Present them with too many options and the chances are you'll confuse even a really attentive patron (plus you're showing zero editing skills or decision-making ability). Show them too few and you're underselling yourself (and you risk leaving out what could be the winning idea). I've often presented what I thought were the best options only to have the client pick something from my reject pile. So my suggestion is show everything but clearly divide them into an A and a B list, with an explanation of each. This is particularly important if you're responding by email where you don't have the benefit of introducing your work in person.

Stand up for yourself

If you get the opportunity to present your own work to clients I suggest you take it. Don't leave it to others to do the talking for you. Even if they're polished and you're rough-hewn, with fingers of butter, fists of ham and feet of clay when it comes

> deal with clients as an equal, not a supplicant

to presenting, it shows you care and that you take ownership for your output. Plus it's a good occasion to get first-hand feedback that could result in a far better piece of work. The trick here is don't try too hard. Deal with your audience as an equal, not a supplicant. Naturally you want them to like your work, but it's not the end of the world if they don't (even if it is). As any hot date will tell you, neediness isn't very appealing.

In a nutshell:

- Use the brief to settle disputes.
- Ask what's wrong, not what's right.
- Take responsibility for any errors.
- Read between the lines to help them understand themselves.

- Show everything, but divide your work into an A and a B list.
- Present yourself as their equal.

CHAPTER 2

Three key thoughts

So far I've described the role of writer and sketched out what the job of copywriting entails on a day-to-day basis. Now let's go deeper with three key thoughts that crop up again and again throughout this book.

Thought No. 1 – Don't be dull

If there's one thing I want you to take away from *Brilliant Copywriting* it's this: *don't be dull*. If you forget everything else here (and I sincerely hope you don't) then let this one thought remain lodged in your brain.

Boring is bad because *boring doesn't work*. According to Howard Gossage, an iconoclastic copywriter and adman from the 1950s, 'No one reads ads. They read what interests them.' I'm sure that with enough repetition (which usually means money) even the most moribund message will ultimately produce a modest upturn in sales or awareness or whatever. But believe me, most clients would rather outthink their competitors than outspend them, and most copywriters would like to help them do it. And that's where making stuff interesting comes in.

Here's what Bill Bernbach, one of US advertising's most celebrated and influential figures, had to say on the subject:

The truth isn't the truth until people believe you, and they can't believe you if they don't know what you're saying, and they can't know what you're saying if they don't listen to you, and they won't listen to you if you're not interesting, and you won't be interesting unless you say things imaginatively, originally, freshly.

interest is an essential prerequisite for understanding and action

That's pretty much this whole book in a paragraph. Interest is an essential prerequisite for understanding and action. Only when an idea is presented as relevant and engaging will a reader really connect with it. Another Bernbach aphorism was, 'No one is waiting to hear from us.' On the contrary, most people view most marketing messages as unwarranted intrusions into their day and automatically screen them out. Actually, that's putting it politely – many people reserve a particularly potent kind of loathing for the avalanche of ads with which they're assailed on an hourly basis. To get around this, our copy needs to justify its existence pretty damn quick. It can work: genuinely effective, interesting, entertaining, dramatic copywriting not only gets people's attention, it can also earn their affection. But *only* if it's interesting.

brilliant tips

Three quick anti-boring techniques

One: The truth sets you free

How do you banish boredom and overcome objections? One way is to be disarmingly honest. Quentin Crisp remarked that, 'Everyone who tells the truth is interesting', presumably because we're so used to being surrounded by routine duplicity and double-talk. Crisp's sentiment is echoed in yet another quote from Bernbach (sorry, he's just so quotable), 'I've got a great gimmick – why don't we tell the truth?' Obviously this advice doesn't cover how to actually express the truth in a way that will attract the right response (that's something we'll look at later), but as a starting point it's hard to beat.

Two: Believe in your subject

It's sometimes said that there are no dull products, only dull copywriters. I'm not sure it's quite that simple, but this does touch on an important anti-dullness technique: if possible, avoid writing about products and services you're not personally interested in. OK, that won't be easy a lot of the time, but some subjects – like sport and music – demand real passion in their writers, and it's very hard to fake that kind of fervour. One agency I worked for had the FA account. I knew nothing about football and cared even less, so I lived in fear of being asked to write about the new England away strip or whatever. Get this stuff wrong and it's easy to sound like someone's dad trying to be down with the kids. So if you've any choice at all, only write about high passion subjects you've some acquaintance with (or at least don't actively dislike).

Three: Life is short, copy is long

As you write I recommend you regularly apply what a friend of mine calls 'the nursing home test'. Here's how it works. Imagine yourself old and frail, spinning out your days in a comfy, overheated nursing home, pleasingly befuddled on prescription drugs and disturbed only by visits from doting grandchildren. As you reflect upon your career as a copywriter, can you honestly say that you're proud of what you've done? I don't mean, 'Did I create great art?' What I'm talking about is making sure each piece of work was as good as it could be under the circumstances of its creation, and that it had a kind of internal honesty that gave it integrity. Try applying the nursing home test to everything you write – it won't all pass, but the more that does, the more you'll create copy that deserves to be called brilliant.

Will you pass the nursing home test?

In a nutshell:

- 'No one reads ads. They read what interests them.'
- 'No one is waiting to hear from us.'
- Try telling the truth.
- Avoid high passion subjects you don't care about.
- Will you be proud of it in 30 years' time?

Thought No. 2 – Write like you speak

We've established that boring is bad. So how can you write brilliant copy that's as engaging as it is effective? One of the best ways I've found is to talk to your reader as if they're sitting in front of you. In other words, have a conversation. And that means thinking of them as real, live, human beings.

The idea that a conversational approach to persuasion yields the best results isn't exactly new. Queen Victoria complained that Gladstone talked to her as if he were addressing a public meeting. She preferred Disraeli, who spoke in a less pompous, hectoring tone. My advice is that you follow Disraeli's example, even if your audience is far from regal.

Selling stories

Stories and conversation walk hand in hand down the path of persuasion. Take a look at the websites of a few established copywriters and you'll soon come across phrases like 'corporate storytelling' and 'the power of stories to inspire action'. I think it's easy to overdo this angle but there's no denying that true stories offer tangible proof of abstract claims and work brilliantly as persuaders, particularly when it comes to selling ideas. That's because ideas often need images in the form of stories to make them feel real. Without these images, an idea – even a really good one – can appear hopelessly abstract. The question becomes, how can you relate your idea to something the audience already understands? And the easiest way is, you guessed it, with a story.

> true stories offer tangible proof of abstract claims and work brilliantly as persuaders

Presenting information in story form also helps make it personal as our minds automatically try to make sense of what we're hearing and apply it to our own lives. These stories can be surprisingly short – it's not the volume of information that stirs the soul; it's interest and relevance. Nor do they need much in the way of context – if the story is embedded within another body of text in a way that doesn't draw attention to itself then it's likely to be even more effective (something Jesus understood). And don't worry too much about detail – the audience will happily fill in the blanks without realising they're doing it.

brilliant tip

Stories really come into their own when you're trying to illustrate a company's values in action. Here the best approach is to tell your readers a micro story that makes your point and then let them draw their own conclusions. I recently wrote a piece for a major retailer that illustrated their value of 'can do' with a story about how branch staff heroically beat back severe flooding to stay open when all around them shut up shop. Another way to put this is 'don't claim, demonstrate'. Stories are a great way to bring this rather abstract piece of advice down to earth. And remember, an organisation can have multiple micro stories – you can tell the story of the need they fulfil (in other words, their market story), how they fulfil it (their product or service story), and what's so special about that particular organisation (probably a people story).

One word of advice – in order to work, a story needs a bit of bite. Typically it must resolve some antagonism. It's often said that 'conflict is the essence of drama', and it's much the same with corporate stories. Dig deep into your client's world and you might be surprised what comes up.

The art of conversation

don't write anything you wouldn't say in person

Stories dovetail nicely with the idea of copywriting as a kind of conversation with the reader. That means adopting some of the strategies of spoken, rather than written, English. In practice that means don't write anything you wouldn't say in person. I'm not suggesting you throw all formality to the wind, simply that the warmth and casualness characteristic of conversations has a much wider application than you might think.

Admittedly we're not always expert conversationalists – as Henry Miller said, 'We don't talk, we bludgeon one another with facts and theories gleaned from cursory readings of newspapers and magazines' (and for copywriters he could have added websites, last year's annual report, brochures, factory visits, interviews and so on). Nevertheless it's an effort worth making, for the more interesting the conversation, the more likely your audience is to engage and the more successful the outcome is likely to be.

So a big part of brilliant copywriting is writing as if you're having an informal, informative conversation with your reader. Tell them the basics, anticipate their questions then give them something extra. Most of all, make it real and make it *interesting*. Good conversations invite listening, and interest is a key part of that process. Conversation allows you to talk about a product or whatever in a way that overcomes resistance because people are predisposed to listen (providing the conversation's any good, of course). And if that happens you're well on your way to a sale.

> the more interesting the conversation, the more likely your audience is to engage

In a nutshell:

- True stories offer tangible proof of abstract claims and work brilliantly as persuaders.
- Think of copywriting as a kind of informal, informative conversation with the reader.

Thought No. 3 – Believe you're brilliant

As a copywriter you're paid to create on demand. Sitting around waiting for the muse to show up isn't really an option, so all brilliant copywriters need a few tricks up their sleeves to help them have great ideas as and when they need them. I've heard this sort of no-nonsense professionalism described as 'blue-collar writing', and it's something I'm very proud to be associated with. No technique for producing ideas can guarantee results, but this section should tip the odds in your favour.

Yes you can

The most important thing I can tell you about having ideas is to *believe you can do it*. I realise that sounds like a piece of sick-making self-help nonsense, but you have to believe or you might as well give up and get a nice job at your local library. Or to put it another way, whether you think you can have great ideas or you can't, you're probably right. Confidence is everything. If you believe, it will somehow happen. That's because *the solution exists*. No matter how impossible the problem sounds, no matter how difficult the brief, it has an answer. In fact it has an infinite number of answers. And the first step to finding them is knowing they're out there.

Look for connections

> creating new combinations of old elements depends on your ability to see relationships between unconnected ideas

Creativity is an act of association – it's about taking two unconnected ideas and finding a way of bringing them together to produce a new, third idea that somehow means more than its constituent parts. Creating new combinations of old elements depends on your ability to see relationships between unconnected ideas. It's this ability to spot a promising link that is the hallmark of creativity and – by extension – brilliant copywriting.

So what do you actually *do*?

The first thing is to force yourself to think that what you have to do is easy. Start thinking of yourself as someone who *can* create on demand. No matter how hateful it sounds, you will improve your problem-solving ability by improving your self-image. How you think about yourself will directly determine your creative success rate. In short, you need to think like someone who has lots of brilliant ideas in order to become that person.

To get started, have a loose discussion in your head (or out loud if you've the opportunity) about what you're trying to say. What's the biggest single problem here? Can you state it clearly in one sentence? If not, keep redefining the problem until you can. This is important: how can you expect to solve what you can't describe? Once something decent emerges, write it down and use that as a starting point. I tend to scribble it down in the middle of a big sheet of paper and then hang associated ideas off this central thought in a rough mind map. If these newly added ideas naturally cluster together away from the centre of the sheet of paper then that might suggest my original insight was wrong and I need to look at everything again.

I might then try drawing connections between related ideas – the more lines a particular idea attracts, the more important it must be. Once I've made all the connections I can, I should be able to see which ideas are the real winners. If I'm doing something longer and more detailed, I can then number these ideas to provide a rough sequence for the whole piece. Now it'll practically write itself.

brilliant tip

Talk to yourself (or a sympathetic other) to clarify the problem. Feel free to ramble as the mood takes you. Once a vaguely promising idea emerges, write it down quickly. Add associated ideas and connect these to the central idea and each other as appropriate – the more connections to a particular idea, the more significant it is. When you're finished making connections try numbering the main nodes, and there's your rough outline.

Getting unstuck

If it's not happening try going for a walk – getting out into the fresh air really does help. While he was writing *On the Origin of Species*, Charles Darwin had a special path (the 'Sand-walk') constructed at his house in Kent for just this purpose. Another technique recommended by many of my interviewees later in this book is engaging the other half of your brain, or at least doing something radically different. Read some poetry, look at a design blog, listen to The Ramones at top volume, do all three – whatever it takes to get your mind off the problem. Other techniques I've found useful include changing location (try a meeting room or the kitchen table instead of your desk), changing tools (computer for pencil or vice versa), automatic writing (keep scribbling about your subject non-stop for a few minutes without hesitation, repetition or deviation) and my old favourite, staring into space.

A technique for producing ideas

For my money the all-time top text on creativity was written in 1965 by a gentleman named James Webb Young, a J. Walter Thompson account executive. It's called, helpfully enough, *A Technique for Producing Ideas* and it divides the creative process into five stages. So universally applicable and effective is his approach that it's worth describing in some detail. I'm not exaggerating when I say that getting to grips with what follows is half the battle of brilliant thinking in any field.

The first stage is to gather your raw material, both specific and general. By specific I mean anything relating to the product or service you're writing about. By general I mean whatever enriches your mind, no matter how apparently off the topic it might seem. As Young wrote, 'It is with the advertising man as it is with the cow: no browsing, no milk.' A similar sentiment was expressed by designer Adrian Shaughnessy when he wrote, 'Cultural awareness, backed up by targeted research, is the high octane fuel that drives great ideas.'

The second stage is a process of mental digestion. You need to masticate the raw material you've collected. Examine the facts from unusual angles, looking for any sort of pattern or fit. Don't read them too directly – it's more a process of listening than looking. But don't scrimp here – you need to thoroughly engage with your texts to do them justice.

If you've undertaken this stage with real diligence then two things are likely to happen: first some initial, half-formed ideas will appear out of nowhere (grab them while they're hot – they could be your answer) and secondly you'll become tired, angry and utterly fed up with the whole wretched process. Excellent – you're making progress.

The third stage is to forget the whole thing. Go away. Do something else. Distract yourself. The best way to do that is to turn to whatever stimulates your imagination and emotions. That could be music, film, art or whatever. Just not bloody copywriting. The aim here is to let all those ideas you swallowed in stage two really ferment in your subconscious. This is my favourite stage.

The fourth stage is the Eureka moment. Out of nowhere ideas will appear. Once you've stopped straining for them and gone through a suitable period of rest and relaxation, the magic *will* happen. They'll be raw and rude, but these ideas will be the genuine product of the material you consumed in stages one and two. Collect them all.

The fifth stage is the patient work needed to make these ideas fit your exact problem. This is a process of tuning, refinement and distillation. As Young points out, 'Good ideas have a self-expanding quality – possibilities in them come to light under further examination.' This stage requires patience and hard work. If the previous four stages are about producing the oft-quoted 1 per cent inspiration, then this stage is about the 99 per cent perspiration part.

your first idea is often your best

I'd like to add an addendum to Young's masterly technique: your first idea is often your best. It's the freshest, the thing that sprang to mind before you filled your head with too much clutter. Don't neglect your other ideas, but treasure that first response. And lastly, when the ideas are coming, when you're in what athletes call 'the zone' where everything comes together, action is effortless and you can do no wrong, *don't stop*. It's tempting to think, 'I've cracked it, time for a coffee'. Try to resist this urge until you are genuinely dry.

In a nutshell:

● Nurture a blue-collar approach to writing – it's a job.

● Force yourself to believe you can do it.

● Look for a relationship between unconnected ideas.

● Talk to yourself to clarify the problem.

● If you're stuck, engage the other side of your brain.

CHAPTER 3

Brands and
tone of voice

A bit about brands

As a copywriter you're going to hear plenty about brands and the related issue of tone of voice. Indeed, 'brand' is one of the business world's most overused and abused words. According to some pundits practically everything is a brand. I'm not so sure. For a person, place, product or service to qualify as a brand it needs some sort of emotional aura, something for its audience to *get excited about*, something to love. Mere existence is not enough.

In spite of this, many organisations still commission detailed brand development projects that deliver a hierarchy of elements starting with some form of big idea, followed by vision, values, personality and so on. You often see this represented visually as a pyramid or a crude, playschool-style house. Despite working on many such branding projects for many different clients, I've yet to see one of these carefully constructed confections achieve anything useful, at least from a copywriting perspective. All too often this work is of no benefit to anyone except the organisation paid to develop it. Radical simplicity – and a hefty dose of honesty – are the answer. In fact to write for a brand, a copywriter usually needs to understand just two components: the big idea and the brand personality. Everything else is a distraction.

As you might expect, a big idea is a one-word or one-sentence description of what a brand is really about when all the hoopla is stripped away. Why is a big idea so important? Well, many organisations offer essentially similar – some would say inter-changeable – products and services (so-called parity products). In this situation the main thing that distinguishes successful companies from their lesser rivals is their worldview, their atti-tude, their special way of doing things. In other words, the big idea that captures and expresses their point of difference. This emotional logic is expressed along the lines of 'I like what you stand for/the way you do business/the way you make me feel'. This matters because companies that stand for something – as captured in their big idea – tend to stand out. And standing out is a big part of what branding is all about.

I'm a big fan of big ideas – they help me do my job. Without a strong big idea, a brand becomes a rudderless Ship of Uncertainty dashed against the savage Rocks of Anonymity in the stormy Ocean of Meaning. Not a good place to be. Asking 'What's the big idea here?' is always a useful question in the early stages of a project. Remember, your role is to create difference and interest in the face of sameness. That means animating the big idea in a way that creates a whole emotional world your reader can buy into (literally as well as metaphorically).

brilliant tip

According to David Ogilvy the way to recognise a truly big idea is to ask yourself:

● Did it make me gasp?
● Do I wish I'd thought of it myself?
● Is it unique?
● Does it fit the strategy to perfection?
● Could it be used for 30 years?

I particularly like the one about wishing I'd thought of it myself; in many ways that's the acid test of any piece of work.

In contrast, a brand's personality is usually defined in a series of adjectives – 'thoughtful', 'dynamic', 'rigorous' and so on. Brand guidelines often bring these to life using example sentences that show too much, too little and just enough of the requisite quality. It's a good system and can be genuinely helpful to us copywriters. A brand's big idea can help steer your general direction; a brand's personality can help steer your tone.

While we're on the subject of personality, it was recently brought to my attention that the words *per sona* mean 'through sound' in Latin and originally referred to the mask worn by Classical actors. With the mask in place their face – and therefore expression – was obscured, so the audience was forced to rely on words alone for their understanding of the action.

It's exactly the same with brands – copywriters can project personality by using language in a way that reveals everything an audience needs to know. How? Well, that's the challenge, but a useful place to start is by picturing someone with the personality traits you're being asked to use. How would they speak? What turns of phrase and word choices would they make? How can the bland adjectives of a brand's personality be transmuted into language that means something? Make your brand a human and you're halfway there. All of which means thinking about tone of voice, and it is to this hottest of hot copywriting topics that we'll turn our attention next.

> make your brand a human and you're halfway there

In a nutshell:

- Copywriters need to understand the big idea and the brand personality.
- Organisations that stand for something – as captured in their big idea – stand out.
- To bring a brand's personality to life, try picturing a person with the personality traits you're being asked to use.

Striking the right tone

Many organisations want a distinctive tone of voice, many copywriters exert themselves creating such a thing, yet few of us ever stop to ask *exactly* what we mean by these three little words. This troubles me, because if we can't define it, we can't deliver it.

Admittedly no single definition is ever going to work in every situation. In fact there are probably as many definitions of 'tone of voice' as there are people using the phrase. Nevertheless, here are three interpretations I've found useful.

Definition No. 1: Content + expression + audience

For most people 'tone of voice' means a writer or speaker's choice of words, phrases, idioms, figures of speech and so on. In other words, the actual language they use.

But tone of voice is just as much about content as it is about expression – in other words, what ideas we choose to include and what we choose to ignore. Tone of voice can't be just *how* we say things, it's got to be about *what* we choose to say as well. You can't have one without the other, and in practice it's almost impossible to separate the two.

Then there's a third factor that influences tone of voice – *the audience*. If the language you use and the ideas you choose to present are wrong for your reader, then all the eloquence in the world won't help. In fact, knowledge of your audience should be the *first* thing you think of when trying to determine the right language to use and the right content to include. It's our old friend 'remember your reader' cropping up yet again.

brilliant tip

Don't automatically reject jargon. While it's generally true that brilliant copywriters should speak like normal human beings, it's also true that jargon does have a place. The thing is, jargon *works*, especially as a shorthand between members of any special-interest group. The problem with jargon is that it excludes outsiders (some would say that's half its point). The answer for copywriters is simple: remember your reader. If you're writing for a clearly defined group with their own micro culture, then fill your page with all the technobabble they can take (as long as you're sure they can take it) – in fact you *must* in order to appear credible. But if your words are destined for a wider audience then it's up to you to move heaven and earth to translate as appropriate. The point is to keep your writing relevant for your audience.

Definition No. 2: Personality in print

It might seem an odd thing to say, but reading is really an act of listening. When you read a novel the voice in your head is telling you the story; when you read a poem that voice is telling you what the poet is feeling. In fact, think of any great author – one of the things that makes them great is the voice you hear when you read their words. What you're experiencing is their ability to project a personality onto the page.

> reading is really an act of listening

So one useful way to describe tone of voice is 'personality in print' (or to put it another way, 'expression as differentiation'). When you're in discussions with a new client, particularly if you're not familiar with their brand, one of the first questions to ask is, 'Whose voice should we hear when we read the words?' If they can't tell you (and they probably can't) then you need to somehow reach a consensus, probably by referring to their brand personality in some way. This is important stuff – unless you know the voice you're aiming to emulate (it might be a founder, the current CEO, a media figure or some composite) you can't possibly write in a way that's right for the brand. As an example let me mention some writing I undertook for a charity that oversees some of London's most historic buildings. After much heated debate we ended up with a tone of voice we described as 'part Tony Robinson, part David Starkey'. With that agreed it was simple to write in the right tone. It also gave me a semi-objective yardstick against which to measure our efforts. In many ways 'Whose voice should we hear?' is the most important question a copywriter can ask a client.

Definition No. 3: Everything you don't have to say

Another definition of tone of voice I like very much is 'everything you don't have to say to get your message across, but probably should'. If that seems unwieldy just think it over for a moment. Far from being unimportant, the little things you don't have to say – the turns of phrase, rhetorical devices, idiosyncratic twists, unusual word choices and so on – are essential. The extra material that surrounds your core message is the thing that really differentiates what you say. It's this that gives language its power.

> in writing, personality comes from all the unnecessary extras

In fact the more a copywriter simplifies a text by cutting away anything that is surplus to the requirements of basic intelligibility, the more they reduce the possibility of creating an original tone of voice, simply because there are fewer words to play with. In writing, as in life, personality comes from all the unnecessary extras.

This is potentially dangerous territory. Elsewhere in this book I advise you that when it comes to copywriting, less is almost always more. And so it is ... except when it isn't. Generally speaking, a piece of writing should be as short and pithy as possible. However, if you take this too far you end up with over-impacted language that can be off-putting to readers because it packs too much meaning into too small a space. As John Simmons puts it, 'Language is rarely at its best when it is at its most direct.'

As proof, consider the following fictional PowerPoint slide. You may recognise its content:

My Dream/American Dream

Main points:
- Live out our national creed
- Advent of brotherhood in Georgia
- Transformation of Mississippi

Criteria of personal judgement:
- Colour of skin? NO
- Content of character? YES

It is, of course, the Rev Dr Martin Luther King, Jr, taken from a speech delivered to around 250,000 people on 28 August 1963 at the Lincoln Memorial, Washington DC during the March on Washington for Jobs and Freedom. Here's the original:

I still have a dream. It is a dream deeply rooted in the American dream.

I have a dream that one day this nation will rise up and live out the true meaning of its creed: 'We hold these truths to be self-evident, that all men are created equal.'

I have a dream that one day on the red hills of Georgia, the sons of former slaves and the sons of former slave owners will be able to sit down together at the table of brotherhood.

I have a dream that one day even the state of Mississippi, a state sweltering with the heat of injustice, sweltering with the heat of oppression, will be transformed into an oasis of freedom and justice.

I have a dream that my four little children will one day live in a nation where they will not be judged by the colour of their skin but by the content of their character.

The longer, more discursive, more human, version has infinitely more power. The point is that tone of voice – in the form of all the things you don't strictly have to say – gives words their impact.

In a nutshell:

- Tone of voice is content + expression + audience.
- It is personality in print.
- It is everything you don't have to say to get your message across, but probably should.

Clarity, not simplicity

Over the last few sections I've made the point that a conversational approach to language is an important part of brilliant copywriting. That's because conversations communicate personality, which brings the brand to life and gives the customer something to like. Most conversations tend to ramble a bit, which I think is a very good thing. It relates directly to my definition of tone of voice as 'everything you don't have to say but probably should'. The little extras make all the difference, and that has implications for any copywriter seeking simplicity.

You don't have to go far in the world of branding, advertising and design to hear a clamour of voices expounding the power of simplicity. The assertion that less is more has become an article of faith among creatives everywhere, but is this always right? My view is that while simplicity is good, clarity is far, far better. The information expert Richard Saul Wurman makes the point that while clarity is an essential prerequisite for understanding, simplicity often means taking away the very bits that made the message interesting in the first place. The line between simplicity and simplistic is precariously fine; crossing it can have disastrous consequences for understanding.

While the pursuit of clarity may deliver simplicity as a by-product, this isn't always the case – it all depends on your audience and context. When I'm writing I'm constantly torn between the (perfectly laudable) desire to simplify, and an awareness of the complexity with which my audience actually talks. By substituting clarity for simplicity the problem dissolves. Paradoxically, an emphasis on clarity may mean leaving some aspects of your message open to interpretation, ready to be completed in the mind of your reader. A clear but open-ended message can sometimes make a lot more sense than either a simpler version that leaves out essential details, or a lengthy explanation that tries to nail down every semantic loose end, boring its audience to death along the way. If the message is right for the audience, and the audience is right for the message, they'll get it – simple as that. It's a powerful realisation, but like many such realisations, frequently forgotten.

> if the message is right for the audience, they'll get it

Exactly what 'Just Do It' has to do with fashion gear presented as sportswear isn't clear, but it works. Put the right cues in front of the right audience and meaning will detonate in their minds.

In a nutshell:

- Clarity is an essential prerequisite for understanding.
- Simplicity often means taking away the very bits that made the message interesting in the first place.
- Don't be afraid to use a few extra words to create the right tone.

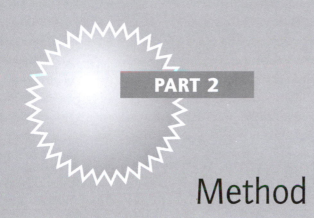

PART 2

Method

et's recap: in Part 1 we established that it's bad to be boring, a conversational approach can create interest, tone of voice is about the individual voice you heard when you read what's on the page, and a few extra words can make all the difference when it comes to creating brilliant copy. In short we've talked over some of the background to being a brilliant copywriter. Now let's zoom in a bit and look at the practice of writing.

I've divided the writing process into three stages: before, during and after. That's an over-simplification of course – in real life the boundaries are blurred as the various stages overlap, repeat and run in parallel. Still, it's a useful and intuitive way of bringing clarity to a confused and essentially subconscious process. So don't worry too much about how I've chosen to structure this part – just take what you need and apply it where and when you need it.

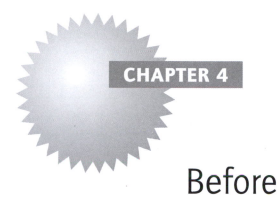

CHAPTER 4

Before

'Proper planning prevents piss-poor performance', as they say in the Army, a sentiment equally true in the less martial world of marketing copywriting.

The brief

The first step is to find out what I'm supposed to be doing, and that starts with the brief. In theory this will be a well thought-out, professional document that contains enough information to make the job burst into life before my eyes. In reality it's likely to be a jumble of irrelevant detail cobbled together by an under-motivated minion and presented via a vaguely worded email or hurried conversation, possibly involving a crackly conference call.

If the brief is anything less than 100 per cent clear – and the chances are it will be – then my first job is to take a step back and get all the information I need to move on. This calls for delicacy and persistence, but it's essential for the project's overall success. And what constitutes a good brief? That depends on the job, but I like to see clear, one-sentence definitions of the task, the reader and the overall goal, along with sufficient background information to at least start me thinking.

Once I've acquired a workable brief I like to read it first quickly, then carefully. In the first pass I just want to get a feel for the project; in the second I underline key words in an effort to get a full understanding of what's required. Then I like to rewrite the key points of the brief (while checking I've not missed anything) to fully excise the fluff that seems to clutter up such documents. The really important thing is to find the verbs, because they signpost what I'm being asked to do. By highlighting these I'm well on the way to understanding my mission.

All this checking might seem unnecessary but I can't over-emphasise how important it is to *give them what they ask for* (in other words, respond to the brief as it actually is) rather than give them what I *think* they ask for. It's surprisingly easy to make this mistake but equally easy to avoid – just read the brief very, very carefully and if you're in any doubt about any of it seek clarification. Far better to ask a few extra questions at this stage than cock up the whole project because you went off in the wrong direction.

brilliant tip

The tighter the brief, the more freedom you'll have, for the curious reason that your limitations set you free. This comes up again and again in the interviews at the end of this book. Being told to 'write what you like' sends otherwise effective writers into spasms of self-doubt, so learn to love your limitations and all will be well.

In a nutshell:

- If in the slightest doubt, get clarification.
- Find the verbs. They signpost what you're being asked to do.
- Respond to the brief as it actually is, not as you'd like it to be.
- The tighter the brief, the more freedom you'll have.

The reader

Once I've got a handle on the overall job, I need to know who'll read my words. The brief should spell this out, but if it doesn't I need to make enquiries. As a copywriter it's up to me to take responsibility and really nail my reader. This quote from adman Claude Hopkins – written in 1908 – says it all:

Don't think of people in the mass. That gives you a blurred view. Think of a typical individual man or woman who is likely to want what you sell. The advertising man studies the consumer: He tries to put himself in the position of the buyer.

Hopkins' point is that copywriting is about establishing a one-to-one connection with the reader. To do that I need to speak directly to their emotions in some way, and to do that I need to get inside their head. It's not about what I want to say, it's about what the reader wants to hear. Think of it as a deal along the lines of 'I'll keep reading provided you keep me interested'. I try to make sure I keep my side of the bargain.

> it's not about what I want to say, it's about what the reader wants to hear

In fact, if a piece of copy is anything but brilliant then the chances are that something very simple has happened – the writer has forgotten the reader. As a copywriter my sole purpose is to communicate effectively in an effort to sell my wares – anything less should send me into a frenzy of rewriting. So I try to visualise the person I want to write to and speak to them as if they're sitting across the table from me. I try to picture them in as much detail as I can muster. What's on their mind? What makes them excited? Fearful? Bored? How can I get around the 'So what' question? How can I answer the 'What's in it for me?' enquiry? Only by doing this can I be sure I won't solve the wrong problem when I come to write.

This makes sense when you think about it – there's no such thing as 'readers'; instead my audience is made up of living, breathing individuals linked only by the slenderest potential interest in my subject. If I address them as such I've more chance of getting my message across.

In a nutshell:

● Copywriting is about establishing a one-to-one connection between writer and reader.

● Speak directly to the reader's emotions in some way.

● Talk to them as if they're sitting across the table from you.

● It's not about what I want to say, it's about what the reader wants to hear.

Research

With the broad outline of the job and the fine detail of the reader in place it's time to dig deeper. Brilliant copywriting is largely about the unglamorous practice of preparation, which in turn involves patience and persistence. George Washington said, 'If I had nine hours to chop down a tree, I'd spend six of them sharpening the axe', and every brilliant copywriter needs to acquire the discipline necessary to sharpen their axe before heading off to the woods. Copywriter Ed McCabe put it another way, 'Never go "Ready, fire, aim" – you'll only shoot yourself in the foot'. Wise words.

> brilliant copywriting is largely about the unglamorous practice of preparation

The fact is that if I get the research right my piece will practically write itself. So I start digging. I want to develop a point of view. I want to bring the big idea to life. And as I'm collecting ideas I try to think visually. The chances are that my words will be accompanied by some form of graphic design, so I want to make

A lesson in preparation.

sure the two dovetail. If I'm working with a designer or art director, now's the time to get together. I want to find out what's on their mind and introduce them to any proto-ideas I've come up with.

A simple way to kickstart my research, especially for a large or vague subject, is to do what any cub reporter would do and look for the who, what, where, when, why and how of my subject. That way I'll unearth key facts, and facts have real persuading power. Just as importantly, I need to ask myself what the piece is *really* about. It might be one word; it might be a sentence, but never more. Charles Revlon used to say, 'I make lipstick, but what I really sell is glamour.' So I try to look beyond the obvious to find what I should really be saying.

As the research process progresses I usually build up a library of material, both printed and electronic. Typically only a tiny proportion will be truly relevant to my brief, so I need to patiently pick through, looking for pearls. Earlier I made the point that understanding comes only after assimilation, so it's important I give myself plenty of time to read, digest and understand. As with the brief, I like to read raw material multiple times, looking for different types of relevance. I don't bother trying to understand deeply or connect carefully at this stage – it's more about just grouping together the good stuff.

Along with printed material and web pages I may get the opportunity to interview people from the client organisation. If at all possible I try to arrange one or more such meetings, because they're an incredibly efficient form of research – just being able to ask questions and pursue a particular line of enquiry with a knowledgeable individual knocks other forms of enquiry into a cocked hat. Obviously it's important to get an accurate record of what's said in such meetings. Sure, I could scribble like the wind, but I'd probably miss most of it. And even if I didn't, few people can maintain eye contact, think of the next intelligent thing to say and listen well while writing fourteen to the dozen. Plus, it's polite to listen. People like it. *Clients* like it. The answer to all this is simple: use a voice recorder or Dictaphone.

Few copywriters own one of these little devices, which amazes me. I can't tell you how often recording a meeting has saved my bacon, enabling me to spend precious minutes with the client progressing the project rather than scribbling unintelligible notes. Clients will often tell me exactly what they want – not necessarily in their big speech but while they're shuffling their papers or fiddling with their phone – so I make sure I catch it (I always ask permission to record a meeting, although in my experience very few people object). I always transcribe the recording myself (including all the interviews later in this book) – although excruciating, it concentrates the mind marvellously. The simple act of going over and over the recording often causes an idea to spring forth – so using a Dictaphone can help with inspiration as well as accuracy.

In a nutshell:

- Look for the who, what, where, when, why and how of your subject.
- What's the piece *really* about?
- Give yourself plenty of time to read, digest and understand.
- Interview key people, and use a Dictaphone.

Planning

Once I've gathered enough raw material to feel slightly out of control, I start mapping it out. If I've got a rough idea of my overall argument then I try superimposing my raw material onto this structure to see what works and what doesn't. If I don't have a clue (and I usually don't) then I try writing down the key facts I've uncovered on Post-It notes or an A3 pad, before grouping them in different ways to see what themes emerge. I'll do this a few times to find the optimum organisation. I'll then give names to the groups, order them (again trying a few alternatives) and use that as a basic section plan.

All the time I'm thinking about the reader, the end result and the big idea. I use these as a yardstick to determine whether a particular fact or turn of phrase deserves to be included. I'll then go back and look at my raw material to see where I need more work and if I've missed anything before finessing everything several times until I can't stand the sight of it. Then I'll grit my teeth and go through it once again to make sure I haven't missed anything obvious.

The result should be a pool of relevant information and a more or less detailed structure on which to hang that information. It should reflect the brief and take into account everything I know about the reader, the brand and the desired end result. This approach works whether I'm writing three sentences or 3000 words, although it's on larger jobs that it comes into its own.

brilliant tip

Unusual thinking doesn't tend to happen in usual places. The copywriter Terry Lovelock apparently thought up the line 'Heineken refreshes the parts other beers cannot reach' during a fortnight in a smart hotel in Marrakech, deliberately chosen as a retreat from the distractions of London agency life. I'm not suggesting you need to go quite that far, but the point is sound – sometimes you need a change of scene to give yourself permission to think differently. On a related subject, some people enjoy being driven by deadlines and leave everything until the last moment. Unless you have supreme confidence in your abilities I suggest you don't try this (at least to begin with). Instead I'd start nice and early – you can always slow down later on. Good things happen when you don't push too hard, although I know many writers who take the opposite view.

What it actually looks like depends on how I'm working that particular day, but I tend to do this stage electronically, using different point sizes for different levels of heading, with rough body text bolted on wherever appropriate. It's more than a plan and less than a first draft. What I'm looking for is a growing sense of impatience about the planning process and a desire to start writing. I try to fight this feeling as long as I can, but sooner or later I get to a point when I just know I have enough material, in approximately the right order, to make a decent start with the real thing.

In a nutshell:

● Don't start until you feel slightly out of control.

- Try writing down the key facts on Post-It notes or an A3 pad . . .
- . . . then group them in different ways to see what themes emerge . . .
- . . . then order the themes to create a rough structure.
- The better your plan, the easier your writing.

Originality

Helmut Krone, one of the greatest advertising art directors of all time, said, 'Until you've got a better idea, you copy'. What he was getting at was that you learn a lot by mimicking your heroes. There's nothing wrong, and everything right, in taking a professional cue from those with more knowledge and experience than yourself. If you can find such people, try to establish a connection because their input will accelerate your professional development in a way little else can.

This leads to a larger point concerning the role of originality in copywriting. Basically it isn't as important as you might think. This is commerce, not art. And let's be honest, how many ideas are genuinely original? Very few. Does it matter? Not much. Indeed, it's been said that originality only became 'important' as a result of changes in copyright law during the eighteenth century – in other words originality is more to do with charging royalties than breaking new ground. Whatever the truth, the important thing is to either acknowledge your sources or (as Einstein recommended) successfully disguise them. We all mix, match, cut and paste when it comes to generating our so-called new ideas. That's what creativity is – the bringing together of unrelated ideas and the creation of a previously unimagined connection between them. There's a great quote from film director Jim Jarmusch on this very subject:

> either acknowledge your sources or successfully disguise them

Nothing is original. Steal from anywhere that resonates with inspiration or fuels your imagination. Devour old films, new films, music, books, paintings, photographs, poems, dreams, random conversations, architecture, bridges, street signs, trees, clouds, bodies of water, light and shadows. Select only things to steal from that speak directly to your soul. If you do this, your work (and theft) will be authentic. Authenticity is invaluable; originality is nonexistent. And don't bother concealing your thievery – celebrate it if you feel like it. In any case, always remember what Jean-Luc Godard said: 'It's not where you take things from – it's where you take them to.'

(Reprinted courtesy of MovieMaker Magazine – www.MovieMaker.com)

> the solution to most copywriting problems is common sense expressed with genuine enthusiasm

So to be a brilliant copywriter you don't need to be a brilliantly original thinker; instead you need to be a brilliant collector of other bits of brilliance. The solution to most copywriting problems is common sense expressed with genuine enthusiasm. In the real world, commitment and competence usually trump creativity.

In a nutshell:

- Mimic your heroes.
- Take inspiration from anywhere and everywhere.
- You don't need to be brilliantly original thinker.
- You do need to be a brilliant collector of other bits of brilliance.

CHAPTER 5

During

've read my brief, I've defined my reader, I've done my research and I've cooked up a plan. With any luck I'm now panting with anticipation at the prospect of writing. Here's how I do it.

It's been said that, 'No one writes as well as they'd like, just as well as they can.' How true. However, if you believe that what you write *will* be

> copywriting is largely about giving a damn

read then the chances are you'll write better. Copywriting is largely about giving a damn. That starts with making one last effort every time to really polish your prose, and *that* starts with getting started.

Getting started

Like many copywriters I'm deeply troubled by the thought of settling down to write. I find myself tempted – nay, *driven* – to delay the inevitable with a series of increasingly bizarre pre-writing rituals. This section describes some of the ways I try to get over myself.

Where you invest your attention is what you become

The key to getting anything done is giving it your undivided attention. To put it another way, how I manage my time is neither here nor there, it's what I do in that time – how I direct

don't mistake time spent staring at the screen with progress

my attention – that matters. If I want to get that damn annual report done then the amount of time I spend on it is pretty irrelevant; it's the extent to which I focus on it within that time that will really make a difference. What I'm saying is don't mistake time spent staring at the screen with progress – instead try cultivating a high-focus, short-burst approach to writing.

The thing you're avoiding is the thing you need to do

Displacement activities like cleaning the fridge or shampooing the dog are good, in a strange way. That's because task avoidance helps highlight the task I need to be doing. The chances are that the very thing I'm avoiding is the very thing I should be cracking on with. That's why it's the thing I fear – it carries the most weight. Avoiding it will only make things worse and increase my growing sense of alarm. Of course, tackling thorny problems is never easy, which is why the next suggestion may be of help.

If it's tough, break it up

Some tasks *are* inherently scary – difficult phone calls, important pitches, high-profile pieces of writing for high-pressure clients. The result can be a sort of creative paralysis. To get around this I cut the task into bite-sized pieces I can manage without losing any more hair than I already have. In fact I divide, divide and divide until I've got something I can handle without even a twinge of fear. I then work through the bits in the right order but treating each as a separate job, usually separated by cups of tea and aimless web browsing. That way I can fool my inner coward into thinking all is well. Actually I'm not fooling anyone – using this technique, all *is* well.

Control impulsive distractions

On the subject of web browsing, I don't know about you but the Internet is something of a mixed blessing. On one hand it gives me instant access to a world of interesting stuff, and on the other ... it gives me instant access to a world of interesting stuff. The temptation to waste time in the name of research is immense. Checking email is another black hole into which the day disappears. The solution is in two parts: I try to cultivate a sense of awareness about time wasting and bring myself back to the task in hand once I notice what's going on, and I try to sidestep the whole problem by working in high-concentration bursts with distractions scheduled in between as a sort of reward. If it gets really bad I shut down my browser and email, and labour in splendid isolation.

Do the hardest bit first

If one part of a job looks distinctly harder than the rest, I make completing that my priority. That way I can comfortably dedicate the biggest chunk of time to it without panicking that the deadline is racing towards me like a runaway train. Plus, by the time I've completed it I'll have got into the swing of the whole thing and doing the less tricky parts will feel like a breeze.

Work when you work best

I'm a morning person. If I haven't started by 8.30 I feel dirty and begin to hate myself. I can get quite a bit done before lunch, whereas the period between 12 and 3 is really only good for staring out of the window. Then things pick up again for a burst of semi-decent productivity until I knock off about 6.30. If I have to work late I try to crack on because by 9 in the evening I'm good for nothing. So pay attention to when you're most productive and structure your day accordingly, with admin and so on filling in the downtime.

Do less to do more

Time is not infinite, so to do something really well I need to stop doing other stuff that just gets in the way. That means prioritising on the basis of what really matters (which, incidentally, is almost certainly the thing I'm avoiding most). It also means making space and clearing my schedule. None of that is particularly easy, so if I really do need to clear the decks I start by making a list of everything that's distracting me and work through it, dealing with everything one task at a time.

In a nutshell:

- Attention is the key to getting stuff done.
- The thing you're avoiding is the thing you need to get on with.
- If it's tough, break it up.
- Shut down your browser and email.
- Do the hardest bit first.
- Work when you work best.

First draft

Even if I've done all my research as diligently as possible and put procrastination firmly in its place, there's a good chance my first draft will stink, at least in places. That's fine – the important thing is to get something down. It's far easier and more efficient to work on something that's committed to your computer than to endlessly debate the right approach without getting anywhere.

Choose your tools

If I'm doing headlines, straplines, names or other forms of compact writing, I use a pen and an A3 pad. If it's anything longer I type directly into the laptop. If I'm working with paper I like to divide the sheet into two columns – one for fully formed thoughts, the other for jotting down odd stuff that may be useful later. It's the same on screen – I'll have two Word files open side-by-side, a main document and a scratch pad for natty ideas that may prove helpful, and I just swap stuff between them as appropriate.

Ask – and answer – the right question

One thing I've often noticed as I'm writing is that if I can't find the right answer, I'm probably asking the wrong question. The solution is to revisit the brief and find another angle with more promise. I'll try rephrasing the main point of the brief until I find a chink in the problem's armour, then I really let rip.

> if I can't find the right answer, I'm probably asking the wrong question

A change can be as good as a rest

If I'm stuck I find that changing locations can work wonders – if it's not happening at my desk I might look for an empty meeting room; if I'm working at home I might ditch the office and try the kitchen table. Similarly a change of tools can kickstart my creativity, so if it's not working with Word I might try a stubby pencil and a Moleskine notebook instead. At least that way I can pretend I'm Bruce Chatwin for a few minutes.

Try to have fun – really

If I don't enjoy writing it, the chances are no one will enjoy reading it, so I try to have fun – or at least not feel sorry for myself. That usually involves reminding myself that there are lots of truly terrible jobs out there, and copywriting – for all its apparent *sturm und drang* – isn't one of them. It's a brilliant, creative, hugely privileged thing to be doing. I can't emphasise that enough.

It's a draft, not the final proof

For my first draft I just want to get strong ideas down on the page. I can reorder them, edit them, improve them and so on later. I've also learned that an imperfect full draft is better than two pristine paragraphs, so I get it down, all of it, in the knowledge that I can tidy it up later. The pernicious influence of perfectionism that causes copywriters to rework a tiny part of the whole again and again is really cowardice masquerading as diligence. There is no such thing as perfection, so I try to stop worrying. I always overstuff the first draft in the belief that it's far easier to cross bits out if I don't need them than go back to my source material once the trail has gone cold.

> perfectionism is cowardice masquerading as diligence

In a nutshell:

- Divide your page or screen into two – one for fully formed thoughts, one for rough ideas.
- If you can't find the right answer, you might be asking the wrong question.
- Changing locations can work wonders.
- Reject perfectionism – get it down and tidy it up later.

Making headlines work harder

I think of headlines as the textual equivalent of those first few seconds in front of a potential new boss. I need to make a good impression because the stop/go decision is usually made within moments of that first meeting. It's exactly the same with headlines.

Their job is to do some or all of the following: grab the reader's attention, weed out anyone for whom the body copy has no relevance, give a quick overview of the whole piece and encourage punters to read on. A good headline that manages to do one or

Headlines = instant impact

more of these things means 90 per cent of the selling battle is already won. Incidentally, even if a piece doesn't need a formal headline, jotting down something equivalent is a useful discipline as it forces me to focus my thinking into a pin-sharp phrase.

So without further ado, here are a few top tips for heavenly headlines.

Keep it simple

On the whole I try to use the simplest words I can. Ace copywriter Ed McCabe commented, 'Show me something great and I'll show you a bunch of monosyllables', while Winston Churchill expressed the same point with characteristic brevity, 'Little words move men'. Having said that, it's important to emphasise that the occasional colourful word or phrase has a vital role to play in varying the rhythm and texture of my piece. It's a technique I use all the time to create interest. The longer, more highfalutin alternative isn't necessarily wrong (and indeed might be just right) – it's really a question of what's appropriate for my readers.

brilliant example

Small is beautiful

Try *help* rather than *assist*, *start* instead of *commence*, and *about* in place of *concerning*. Similarly, *use* beats *utilise*, *buy* trumps *purchase*, and *get* wins over *obtain*. Put it all together and you get something like this:

Before

Are there any points on which you require explanation or further particulars?

After

Got a question? Just call.

Before

Repair or replacement of malfunctioning components is free of charge.

After

If something breaks, we'll fix it for free.

Don't automatically dismiss the obvious

I could try being ultra-direct ('Golf sale this way') or I could arouse curiosity ('Did you know . . .'). I could offer a piece of news or provide an answer to a common problem ('How to . . .'). I could present the headline as a question, provided the question is genuine and relevant, which probably means it focuses on my audience's self-interest. There will also be times when I can grab the reader's attention using words like *why, quick, easy, free, save, bargain, results* and so on. These are familiar because they work, so use them if they're relevant.

Give readers a reason why

This was the big insight behind John E. Kennedy's description of copywriting as 'salesmanship in print' I mentioned in the Preface. I've said elsewhere that sales are often made for apparently illogical, emotional reasons; that doesn't mean I should neglect the search for clear, rational reasons for the reader to decide in my favour. The more reasons I can give my reader for doing something, the more chance I've got of getting them to comply.

Use plenty of subheads

The headline/subhead pair is a brilliant combination that I strongly recommend you try. Having two bites at the textual cherry means I can write one line in a straight, informative style, and the other as something more kooky and intriguing. That way I can both explain and draw readers in – a sort of 1 + 1 = 3 thing. It doesn't matter which way round I play it – if the headline has its head in the clouds I just make sure the subhead has its feet on the ground (or vice versa).

---

brilliant example

Headline/subhead pairs

Try combining a straight headline with an intriguing subhead:

Walker Hulme Solicitors

Not the law of averages

or

Market forecast

No crystal ball required

And turning it around, try combining an intriguing headline with a straight subhead:

▶

Are we there already?

The comfiest kid-friendly sleeper seats of any major airline

or

Penzance on your patio

Everything you need for summer fun without leaving home

Write your headlines last

If I'm unclear which angle is the most promising then I write my body copy first and see which theme emerges as the winner. Plus, I'll know far more about what I'm trying to say by then, so I'm twice as likely to get it right. This advice also applies to first sentences and the introduction of longer pieces as well as headlines proper.

Search your body copy for hidden headlines

I've found many of my best headlines lurking in my body copy, so if I'm stuck I go through everything else I've come up with to see if there's a dapper phrase hiding in there somewhere. Another great trick is to consider rejected headlines as signoff lines, or indeed my signoff line as a headline. It's surprisingly easy to get the right sentiment in the wrong place, so try mixing and matching – it's a fast way to reveal a new perspective on a piece.

> it's surprisingly easy to get the right sentiment in the wrong place

Never throw anything away

It's amazing how often an idea that didn't quite make the grade on one job can be revived to work brilliantly in another context. So I store away anything even remotely good. Even if I don't use it intact second time around it may spark some new line of enquiry. Think of it as recycling the world's precious supply of potential headlines.

In a nutshell:

- Little words move men.
- Don't automatically dismiss the obvious.
- Give readers a reason why.
- If the headline has its head in the clouds, make sure the subhead has its feet on the ground (or vice versa).
- Write your headlines last.
- Search your body copy for hidden headlines.

Make your body copy the best it can be

Body copy is where you make your case. If the headline sells, the body copy tells. It's where the information lives, but that makes it sound a bit pedestrian, which isn't the case at all. On the contrary, body copy is where the process of persuasion often takes place.

Advertising has largely abandoned anything that could be called 'long copy'. Today's advertising copywriting is all about compression – a single word or short phrase that interacts with a powerful visual to achieve the ad maker's intention. Yet body copy – *long* body copy – is in rude health in many other areas of copywriting. If you write for brands, design, PR, sales and so on, you need to know how to create a compelling argument that stretches over paragraphs or even pages. Here are my top tricks for brilliant body copy.

> the headline sells, the body copy tells

Stop writing body copy

Instead just write good stuff that people want to read. The American crime writer Elmore Leonard once said, 'If it sounds like writing, I rewrite it.' I try to do the same. I try to find some way to connect the world of the product with the world of people. Mercedes Benz cars aren't about getting from A to B, they're about personal prestige. Apple computers aren't about emails and Word, they're about coolness. Plus, it helps enormously if I can get genuinely excited about my subject. In the end it's about making the truth as interesting as it can be.

Organise your argument

Use the traditional inverted pyramid structure much beloved by journalists. This basically means put my strongest material up top in an effort to capture the reader's attention and keep them with me as the argument unfolds. To do that I need to divide my main message(s) from subordinate or supporting messages and group accordingly. But the important thing is to *start strong*.

Start strong ... then stay strong

> plunging your reader into the middle of an argument is a great way to grab their attention

For a powerful beginning, I might try doing what composer John Cage suggested and *start anywhere* – plunging my reader into the middle of an argument is a great way to grab their attention. It's far better to confuse them for five seconds than it is to bore them for five minutes. If that feels too much I might write a few warm-up sentences (or indeed paragraphs), I just make sure that I then delete or demote those sections. The result is usually a far, far stronger start.

If I'm looking for a more conventional way in, classic opening gambits include stating an offer, making an announcement, telling a story or asking a question. Equally I could use a provocative quote, stress a benefit or identify with a reader's problem. A useful technique is to ask what would make *me* react? Remember, facts persuade more than polished claims. According to Strunk and White's *The Elements of Style*, 'The surest way to arouse and hold the attention of the reader is to be specific, direct and concrete. The greatest writers ... are effective largely because they deal in particulars.' So the more facts the better, and that means serious research.

brilliant example

Strong openings

Imagine being smashed against a vice, dipped in paint thinner and then thrown from a motorcycle at 30 mph. That's how we test our mobile phones.

or

Business is a cruel, shallow money trench where thieves and pimps run free and good men die like dogs. There's also a negative side.

Bring order to your ordering

There are five main ways to organise the ideas thrown up by your research: by location, category, hierarchy, time and the alphabet. Say I'm writing a brochure for a vintage watch retailer. I could organise the description of the watches by country of manufacturer (location), year of manufacture (time), model (category), popularity (hierarchy) or simply in alphabetic order. The method I choose would depend on, and influence, what I wanted to say. Each method yields a subtly different sort of understanding and lends itself to a different kind of information, so try a few options.

Keep it punchy

Dame Barbara Cartland, doyenne of the romantic bodice-ripper and one of the most prolific novelists of all time, used to say 'God give me short sentences'. It's excellent advice. That said, it's also important to vary the pace a bit, so the occasional longer sentence (or very short one) can do wonders for the rhythm of a particular passage.

Benefits, not features

This is a crucial point at the very heart of brilliant copywriting. I don't write about what something is, I write about what it can do for the reader. It's about answering the question, 'What's in it for me?' As I'm researching I usually uncover lots of information about a product's features – what it does, how it works and so on. Turning this raw material into a list of benefits and expressing that in an appealing way is what I spend a good chunk of my writing time doing. The best way is to divide my page into two columns. In one I write a list of features, and in the other I translate these into benefits by asking something along the lines of 'How does that help?' With a reasonably fulsome list of benefits at my fingertips I then rank them according to what I think is their reader appeal, and that's my basic argument in the bag.

brilliant example

Paperclip feature/benefit analysis

Features	Benefits
Unusual shape	Natural spring action holds paper together securely for a tidy desk. Brings order where there was chaos.
All-metal construction	Reusable to keep stationery costs down. Almost unlimited working life.

Various sizes	Pick the right size for your job. Never be embarrassed by paper/paperclip mismatch again.
Flexible capacity	Holds two sheets or 20. It's your choice.
Available in individual boxes	One box will last you months, saving tedious trips to the stationery cupboard.
Also available in bulk boxes	Save £££, reducing operational expenditure and increasing shareholder return.
Can be chained together	Instant office jewellery for that nice girl/boy in Accounts.

I jest but you get the idea.

Rational is good, emotional is better

All that fact-finding and turning features into benefits will (hopefully) unearth rational reasons for my readers to believe what I'm saying. That's good, but as we've established, it's only half the story.

try to write to their hearts because that's where persuasion really takes place

Brilliant copywriting demands that I balance these rational reasons with emotional reasons to buy. It could be the actual language I use or how I zero in on a particular benefit and dramatise it in a way that hits home for the individual reader. It comes back to my point about copy as a conversation, a one-to-one chat that just happens to be all about selling. I try to write to their hearts because that's where persuasion really takes place – if a reader can hear a human voice in my words then I've massively increased my chances of being listened to.

Let the reader find their own meaning

Sometimes I like to leave the conclusion dangling before the reader's eyes in anticipation of them doing the final bit of meaning-making for me. I lead them almost all the way, but just leave things ever so slightly unfinished. Asking them to help solve the riddle of its meaning creates empathy (provided it's done well, of course) and involves them in the piece.

Sometimes it's best to tell it straight

I know, I know, this contradicts the last point, but if I've got a great proposition I don't necessarily need to smother it in puns or other comedic shenanigans. I just make sure I *answer the question*. If I judge that a particular copywriting task requires a straight bat (and plenty do) then that's fine. It's about being sensitive to my material, argument and reader.

Introduce structure without being obvious

I once came across a piece of advice from copy god David Abbott along the lines of 'Learn to write a list that doesn't sound like one.' I often find myself needing to present a series of points in my body copy, but using a straightforward list is almost certainly a mistake. I need to be more brilliant than that. So I recommend studying how to use linking words and phrases to maximum effect. Sure, I might write my list straight to start with, but I then go over and over it, softening the edges until it reads without interruption and each sentence flows seamlessly out of the previous one. When I'm done I should have a perfect whole where everything counts and nothing can be removed without upsetting the internal balance of the piece.

▶ brilliant example

A list that isn't

I've turned a bulleted technical datasheet I picked at random into a paragraph without losing any detail:

You can use this ATM just about anywhere. Indoors you can add fingerprint recognition features and a large capacity multi-currency dispenser.

Outdoors you can choose a rugged, rain-repelling finish or a tamper-resistant encrypting keypad. It boasts a range of disability access features like an induction loop and adjustable screen height, and even comes in a drive-up version (perfect for petrol stations) and a mobile option (ideal for festivals, fairs and other outdoor fun).

Apply the time test

If I have the luxury of a generous deadline, I like to write my piece, put it away, then return to it a few hours, days or weeks later. I'll inevitably spot some hideous stylistic blooper that I previously missed. Rudyard Kipling said that when he finished a story he put it in a drawer for a few weeks, then went back through it, crossing out the bits he was most proud of first time around.

Apply the embarrassment test

Earlier I advised you to read your draft aloud to check that it makes sense and that the rhythm and punctuation work as they should. There's another important benefit to this technique – making sure it isn't excruciatingly embarrassing. So I ask myself, 'Would I be happy to read it aloud in front of the client? How about in front of my family or friends?' If anything makes me cringe then I know what to do.

It's got to work

Just because my words read well doesn't automatically mean they actually work. It's easy to style a piece of prose so that it appears to say much more than it delivers. The cure is to ask how something *performs*. And the best people to ask are outsiders. I'm a great believer in the benefit a fresh pair of eyes can bring, especially when it's late or the clock's ticking. So I try to get someone unconnected with the project – usually my long-suffering wife – to give my copy a quick read. Like the time test mentioned above, it's amazingly effective at highlighting hidden horrors.

Look for verbs

Verbs are your friend. That's because actions divide tasks, so if I'm writing anything with the feel of instructions about it, I look for the doing words to help structure my argument. They're particularly useful in telling me where to put sentence and paragraph breaks. Incidentally, if I'm writing real instructions then each one should definitely begin with a verb – that's what an instruction is.

brilliant tip

Here are a few tried and tested ways to kill all interest in a piece. In other words, *how not to do it* – a sort of anti-tip if you like. I'd start by talking endlessly about my client and the minutiae of their world, rather than my readers and what floats their boat. I'd then go on to describe the features of their product or service, never its benefits. I'd make sure I tell readers lots of stuff they either already know or couldn't care less about – better still, both. Along the way I'll try to cram in as many complex, rambling, highly formal sentences as possible, with nested clauses, often placed within further nested clauses, to really make the reader work for the meaning. Then I'll remember all that good stuff they taught me at school about never ending a sentence on a preposition, or using sentence fragments, or starting a sentence with a conjunction. Lastly, I won't worry my pretty head about clichés or banal language – the more the merrier. George Orwell wrote that 'the greatest enemy of clear language is insincerity', and the language of insincerity is cliché. But what did he know?

In a nutshell:

- If it sounds like writing, rewrite it.
- Use the traditional inverted pyramid structure.
- To start strong, start anywhere.
- Rational is good, emotional is better.
- Translate benefits into features.
- Would you be happy to read it aloud in front of the client?

CHAPTER 6

After

Once I've got *something* down, I read it aloud, or at least under my breath. This is one of the best tools in the brilliant copywriter's box, and I commend it to you. No other technique separates the wordy wheat from the textual chaff with such efficiency. I'll instantly see what works and what doesn't, where the punctuation needs alteration, which bits are too long (or too short) and a hundred other things. If it reads like a policeman giving evidence in court or an illiterate seven-year-old then I've got some rewriting to do.

Anyway, with the rough words safely saved I can start to edit. This involves revisiting the techniques introduced in the last section and re-applying them with ever-finer degrees of granularity. It's like a carpenter shaping a piece a wood – initially with a great big saw and latterly with superfine sandpaper. Every adjustment – however slight – brings the piece closer to the divine.

The general idea

The Goldilocks rule

I want to make every word count, and if it doesn't then off it goes. Another way of saying this is, 'if in doubt, chop it out'. That's *not* to say that brilliant copywriting is ultra-compressed – as we've seen, a few extra words can

if in doubt, chop it out

Not too short, not too long, just right.

produce a lot of extra meaning. It's just that good writing contains the *right* number of words – as the saying goes, 'It's not how long you make it, it's how you make it long'. It's the Goldilocks rule – not too short, not too long, but just right.

Put yourself into your work

The editing stage is my last opportunity to increase my personal involvement in the piece. If it moves me, chances are it'll move others. This comes back to the point about copy as a conversation with the reader. If I write from the simple perspective of one person talking to another, then the chances are I'll unthaw the chilly tone of voice that seems to characterise so much corporate communication.

brilliant example

Me to you

Don't say

BigBank has a comprehensive suite of mortgage solutions designed to meet the needs of all our customers.

Do say

We've all sorts of mortgages for all sorts of people. In fact we've probably got one that's just right for you.

In short, use *I, you, us, our, your* and *we*. Don't use 'BigBank's people' or 'BigBank's products' – instead say 'our people' or 'our products' – your reader will know exactly who's talking.

Don't be too hard on yourself

Remember that no copywriter, however brilliant, can do the impossible. If the client hasn't brought some real benefit to whatever it is they're selling, then the writer is reduced to empty phrasemaking. Although fun, this isn't brilliant copywriting. So I try to be realistic about what's possible and don't over-claim.

Creating a messaging toolkit

If I'm writing about a particular product or service on a regular or ongoing basis, it's worth creating what I call a messaging toolkit. This is a list of the main messages the client wants to tell the world, expressed in the right tone of voice and agreed by the client.

It shouldn't be too big or it'll become unwieldy – somewhere between five and ten messages should be about right – two sides of A4 max. Typically each message should be a single sentence or sentence fragment, and each explanation not more than a couple of paragraphs. If I can order them that's great, but it's not what's important. What matters is that by doing a bit of work upfront I create a copy resource I can dip into when time is tight to create production-ready text in minutes rather than hours or days. The crucial point is that the text I come up with is signed off and ready to deploy. Sure, I'll probably need to tweak stuff to make it fit a particular job, but it's a great starting place and a real, practical time saver.

brilliant tip

Messaging toolkit

I recently used this technique on a project for an upmarket estate agent. During the research phase it became clear they had – conveniently – ten key messages, each with a short explanation. By writing these up to a finished standard and getting them signed off I had my toolkit. Then, several weeks later, when I was asked to produce a series of print ads to publicise their business, it was simply a question of dipping into the toolkit to select the phrases and ideas needed. The result was final copy in record time. It's a very slick, very efficient technique I recommend you try.

In a nutshell:

- If in doubt, chop it out.
- It's not how long you make it, it's how you make it long.
- If it moves you, chances are it'll move others.
- Consider creating a messaging toolkit.

Writing for the web

I'm not convinced that writing for the web is *that* different from writing for print – I'm still in the persuasion business and I still have the same 26 letters at my disposal, and so most of the techniques I've just introduced still apply. Having said that, there are *some* differences, and that's what this section is about.

I've included writing for the web in my 'After' chapter because I like to write my first draft in a media-neutral way, and then hone it to suit its final format during the editing stage.

Incidentally, I won't be discussing search engine optimisation (SEO) here, for the simple reason that all through this book I've advised you to write in a way that appeals to your reader. And this reader is – one would hope – a real, live, human being. Writing for a search algorithm is an entirely different ball of wax. Out goes all that stuff about 'don't be dull', 'make it interesting' and so on; in comes 'white hat' SEO techniques, such as repeating key words as often as possible, and 'black hat' methods, using hidden text and links farms. The first is very dull, the second is very naughty.

Plus, the exact nature of the crawler algorithms used by the major search engines changes over time – if I gave you advice today it might be irrelevant in 18 months. SEO isn't even appropriate for many websites – in these situations other web marketing strategies can be much more effective. In short, you're far better off looking online for really specific, really up-to-date information on this darkest of arts.

Back to writing for the web. I think people use the web in three broad ways: as *viewers* who want a fast blast of entertainment (YouTube, etc.), as *seekers* looking for specific information (train times, TV schedules and so on), and as *stayers* who might – possibly – hang around long enough to read my copy. It's this last lot I'm interested in, so to win on the web I need to:

- Cut stuff up
- Write for skimming
- Use lots of links.

Let's go over these in more depth.

Cut stuff up

One of the big differences between paper and pixels is that people read significantly slower on screen. You've probably noticed this yourself – it just doesn't feel right. At this point I *could* say something about cutting the word count mercilessly, but what if I've already done that? A more useful piece of advice is to limit each page to a maximum of around 200 words. Half that amount would be twice as

> limit each page to a maximum of around 200 words. Half that amount would be twice as good

good. Cutting my word count in this way *doesn't* mean ditching important information; instead it means dividing my piece into separate pages – perhaps quite a few. I also need to make sure I end each page with some sort of tiny teaser that helps encourage the reader to carry on reading.

Write for skimming

I don't really read web pages, I *skim* them looking for something interesting. I'm not alone – most readers want to make as little effort as possible to find what they're looking for, so I try to give them a hand.

The web is a supremely active medium – if readers aren't clicking they feel they're not getting anywhere, so I need to write in a way that helps overcome this temptation. The way I do this is to turn up the volume on some of the ideas I've just described. In particular I try to interest them into staying with very high-impact headlines, I use lots of subheadings (far more than I would for print) and I pick out key words in bold. If possible I make those key words tell the main story on their own. Many sources on web writing suggest cutting verbiage ruthlessly and getting to the point as fast as possible, but that's what *all* good writing does.

Use lots of links

This builds on the two previous suggestions. One of the great things about the web is that I can use links to introduce brevity *without sacrificing depth*. I do that by cutting it into chunks and presenting those on separate pages, which means my readers only have to look at the things they're interested in. To get this right I need to work on the structure of my text. It's not enough to just break a long piece (by which I mean anything over

> use links to introduce brevity without sacrificing depth

say 200 words) up into sections and say 'Continued on page 2' – I need to modularise it so that it works well in this new format. To do that I use headlines and/or intros that work in any context so readers get the gist of where they are in the overall flow.

In a nutshell:

- Cut stuff up.
- Write for skimming.
- Use lots of links.

Last but not least

Once I've crafted my copy and I'm broadly happy, it's time to stop and check that I've included all my key messages and that my reader can easily understand what I've written. As I've said, the best way to do this is to read it aloud, backwards, upside down, inside out. Better still, I try to get someone else to read it and comment. It goes without saying that I should subject my piece to a stiff proofread and spellcheck.

As I'm reading I ask myself does it answer the question posed by the brief? Is it relevant to its reader? Does the main text fulfil the promise of the headline? Is it readable, believable, specific and persuasive? Does it flow? Does it include a call to action? And above all, is it somehow interesting?

If the answer to all these is 'yes' then I might be on to something.

brilliant tips

Ten steps to copywriting heaven

1 Know your stuff, or at least have it to hand. Without sufficient raw material you'll soon be reduced to waffle, so do some digging.

2 Find the big idea. Be ruthless. What is it you really want to say? This should be the one thing your audience remembers if they forget everything else. If you've genuinely got several big ideas then consider writing several separate pieces.

3 Write your big idea as a headline. Don't try to show off – just tell it straight, you can always fix it later. Everything builds around this.

4 Briefly develop each point. Use the minimum amount of text possible – any more and it'll just confuse you.

5 Write a conclusion. Don't try to be clever.

6 Write an introduction. Do try to be interesting.

7 Now write your piece, which basically means putting flesh on the bones you've just created using the research you did before you started the planning process. If you've planned properly you'll be done before you know it.

8 Using either a straightforward list or a mind map, write down the main topics you want to cover. Be generous – you can always drop a few later.

9 Shuffle your topics into a coherent argument. Don't just settle on the first structure that emerges – try a few alternatives to check you're right.

10 Write sub-points under each main topic. If you find yourself with more than a few sub-points per topic, consider breaking it in two.

CHAPTER 7

Here's one
I made earlier

ctually, here's five – all real jobs for real clients, although non-disclosure agreements oblige me to change the names to protect the innocent. These examples give a peek into my thinking/writing process and show how it's possible to *Make It Interesting* no matter how unpromising the product or brief. Incidentally, 'peek' is the right word. It's just not possible to comprehensively describe what goes on in my head as I plan, write and edit. Nor can I say with any real confidence, 'Here I'm using technique X, and here it's technique Y'. Copywriting just isn't like that.

Example one – Poster

The job

A poster created as part of a campaign designed to promote a new brand of rum to sophisticated spirit drinkers. The brief made a big thing about how this rum shouldn't exist – at every stage its creator, a gnarled old guy called José Fernandez – had triumphed against the odds. Specifically I had to mention how long this rum has been around (70 years) and somehow antici-pate and overcome objections to the fact that it comes from Venezuela rather than the more usual Caribbean. They also wanted my suggestion for a strapline or signoff.

The piece

I started thinking about José (a real person, although now dead) and what he might have been like. Clearly he'd had to overcome all manner of obstacles to realise his rum-making dreams. So he'd probably become pretty feisty and irascible along the way. Hmmm ... feisty ... rum ... spirits ... *spirited*. Slap on some alliteration and there's my strapline:

Spirited stuff

Now for the body copy. First I need to set the scene, so I go back in time as a way of establishing credibility:

Over seventy years ago Señor José Fernandez, the son of a humble Venezuelan fisherman, set out to distil the finest rum.

Now I need some drama, so I want to mention the fact that José had to overcome many difficulties:

Naturally his plan was met with derision on all sides.

The 'naturally' bit hopefully creates a bit of interest along the lines of 'Why "naturally"? I'd like to know'. Next I use a classic rhetorical structure called 'the rule of three' ('I came, I saw, I conquered') to list the main difficulties José faced. These sentences all start in the same way ('They said') – the result is a neat, parallel structure that builds in intensity. I try to use these to highlight some appealing feature of the product or brand:

They said, 'José, come to your senses! Rum comes from the Caribbean, not Latin America.' 'HA!' I said, 'It comes from wherever it is made with love.' They said, 'The heat of Venezuela will ruin your rum'; I showed it produces an intensely rich and complex spirit that is a wonder to drink. They said, 'Rum comes from old, well-established names, not upstarts without experience'; but now my shelves they groan with trophies!

Now it's time to build up to the signoff line by really hamming up the spirited angle with a brief homage to Monty Python's *The Life of Brian* ('My legs are grey, my eyes are old and bent'):

I am an old man, my eyes are dim and my time is past, yet to my critics I say, 'I proved you wrong, you sons of dogs and donkeys!'

Finally, a headline. I've done a 'rule of three' thing once, so I try it again based on the idea of José having the last laugh:

They laughed at my rum. They laughed at me. They're not laughing any more.

Nice and balanced, and in keeping with the body text. A spot of polishing and I'm done.

Example two – Direct mailer

The job

A straightforward low-budget mailer for a repro facilities management company. Its purpose was to get trade customers to renew their service agreement. The brief emphasised how the protection offered by such agreements can save customers serious money and thankfully came with plenty of relevant info attached.

The piece

I started thinking about 'protection' and – shame on me – thought it might be fun to treat the whole thing as some sort of public health leaflet on erectile dysfunction, incontinence and STD awareness. Don't ask. A spot of Internet research on the type of language used in such things gave me the title:

Are you properly protected?

Your most intimate questions answered

By A Doctor MD

Then it's on with the pastiche. First I need to somehow explain what the mailer is all about. Naturally I turn to a pun on VD (I know, I know):

In our practice we see many organisations suffering from PD (or to give it its scientific name, Printer Dysfunction*). Its distressing and, for many organisations, hard to talk about. The cause is always the same: problems with colour printing.*

The brief makes it clear that I need to mention the twin benefits of safety and saving, so I need to find a way of saying that while maintaining the parody:

Sufferers torment themselves with questions like 'Are my printers properly protected?' 'Am I as safe as I could be?' And most painful of all, 'Why does my wallet ache every time I print?'

So far I've teed up the problem, now I need to introduce the solution. That's in two parts – a colour laser copier and a service agreement, so I treat them consecutively, addressing the cost issue directly with a figure (remember, facts persuade):

Luckily our scientists have come up with a solution – a colour laser copier combined with our unique Premier service agreement. You may be thinking, 'But Doctor, won't a colour laser copier be expensive?' Dear me, no – print for print you could save up to 35% compared to a standard printer. What's more, our colour protection scheme is so comfortable you'll hardly know it's there. All you'll feel is the confidence and reassurance that comes from top-notch professional support.

Finally it's time for a bit more mock-medical chicanery followed by a call to action:

With our help you'll soon be performing without problems. Call now on 01234 5678910 for a consultation – you'll be glad you did.

Example three – Comps slip

The job

A design studio I occasionally work for asked me to come up with something more interesting than the usual 'With compliments' for their new stationery. It doesn't get much simpler than that. But as anyone who's tackled similar jobs will tell you, simple can be downright hard. Plus, they wanted the job doing in an hour – well, I like a challenge.

The piece

Straight away I knew I had to do something that twisted the idea of compliments. But no matter how incandescently brilliant my writing, it's still just a comps slip. Once someone's seen it the game's up. How could I increase its shelf life? Then I thought, instead of one design why not have several? That way recipients get something different each time and it becomes something small but significant to look forward to rather than just throw in the bin.

My first thought involved a play on the idea of compliments as praise or admiration. I came up with lines for a series of five slips:

Nice shoes

Looking good

Is that a new tie/shirt/haircut (delete as applicable)?

Have you started working out?

Can I just say you're looking particularly lovely today?

To make the joke work and to signal that each example is one in a series, I add the sign-off 'Another compliments slip from [company name]'.

It's always good to give clients a couple of options, so next I think about famous quotations on the theme of compliments (quotes can be great catalysts). A spot of research revealed:

Your eyes shine like the pants of my blue serge suit.

Groucho Marx

If you can't get a compliment any other way, pay yourself one.

Mark Twain

I have been complimented many times and they always embarrass me; I always feel that they have not said enough.

Mark Twain

Women are never disarmed by compliments. Men always are. That is the difference between the two sexes.

Oscar Wilde

I particularly love Groucho's quote. Again I use the same sign-off line to nail the gag and indicate it's part of a series, and it's job done for now. Time to put it in front of the client and let the horse-trading begin.

Example four – Press ad

The job

Come up with a series of flexible, interchangeable headlines and body sentences to promote a range of budget bedding. The brief made it clear that 'everything had to work with everything' – in other words all the headlines had to work with all the body copy. It also established that the illustrations used in the ads – featuring a soft focus, romantic novel look – were the hero, and whatever I came up with had to work perfectly with them.

The piece

OK, bedding. Sheets, pillows, bedsets . . . that sort of thing. The brief made some mention of their quality feel despite their budget prices, so there might be something in the tactile aspect of the product. Something on the theme of touch might work. And if it's all a bit romantic and starry-eyed perhaps 'touch' could become 'caress', 'sensation' or 'pleasure'. Come to think of it, the whole idiom of romantic novels could be promising territory. So after a quick trip to my local Oxfam for research into the Mills and Boon house style I come up with a potential headline:

One touch was enough

I'm on to something, so I press on:

Yield to me
Gently my darling
Promise me silk
A stolen caress
Smooth as the starlight
Satin nights
A feeling most rare
The sweetest sensation
Only for her pleasure

The last one's a bit racy but they're all worth showing to the client. Now for the body copy. I've found my voice so now I just need to introduce some product features. The brief is light on detail so I have to make the most of the least. Again I'll do a series that I can mix and match with the headlines. After I've messed about for a bit, a two-sentence format starts to feel right:

As she fled into her bedroom Angelica realised she had crossed swords with someone she could not rule. 'I am no man's slave!' she sobbed as she buried her head in the softly yielding mixed fibre pillow (two for £6.99).

Having established a way into the problem I can cover all the points raised in the brief in a methodical fashion:

With her fingers brushing the delicate fabric of the genuine down-filled double duvet, Francesca asked herself how something as gentle as her love for the dashing Count Gustav could arouse such blazing passion.

His touch sent shivers down her spine. 'Why do you torment me so?' she cried, as the delicate lace of the 100% cotton bedset with matching slip pillowcases moistened with her tears.

Her indifferent heart had never been touched! But as they fell upon the 50% polycotton/silk mix bedset something stirred in her soul that no man had ever glimpsed.

'You little fool!' he blazed. 'Don't you know a buccaneer's man-o-war is no place for a lady?' But as her eyes lingered upon his waterproof mattress protector he felt his heart slowly melting.

Again the last is perhaps a little cheeky but certainly worth showing the client, if only to prove I haven't been staring out of the window all day.

Example five – product leaflet

The job

A new brand of luxury espresso needed a small-format leaflet to go inside a smart tin containing their bagged-up coffee beans. I didn't have a brief as such, just a couple of fairly detailed interviews with the CEO and his chief taster and some verbal instructions. What came out of that was a number of key words ('smooth', unique', 'quality', 'excellence', 'obsession' – you get the picture) and lots of background bumph about estate terroir, roasting and so on. It was all good stuff but slightly unfocused and underwhelming – a problem for a drink positioned (and priced) at the super-premium end of the market.

The piece

Reading through the raw material I notice that the taster used the phrase 'the secret of our success' several times. Ah, so they have *multiple* secrets of their success. That's promising – it suggests knowledge, rigour and an agreeable openness about what makes their product so special. Talking about 'the secrets of our success' in the plural is also a slightly quirky thing to say that might snag people's attention. What if I made each 'secret' a spread in the final booklet? Then I could scour the raw material for substantiating facts for each secret and tell a story based around those. I start compiling a list of so-called secrets directly out of my interview transcripts, and after giving them a bit of TLC I end up with:

Secret No. 1 – Quality first. And middle. And last.
Secret No. 2 – Accept no limitation.
Secret No. 3 – Whatever you do, do it with passion.
Secret No. 4 – It's all about the experience.
Secret No. 5 – There's no big secret.

I especially like the last given that I've just been banging on about secrets. That's a good start – now I have a structure for the piece. Next I comb my raw material for substantiating facts, phrases and ideas, grouping these under each secret. Soon I'm ready to compile my notes and convert them into sentence form:

Secret No. 1 – Quality first. And middle. And last.

Our aim is simple: to create the best coffee in the world. This obsession with excellence at any price means [name] is unashamedly for the few. We create our coffee for a very particular audience with a very particular lifestyle – international, cultured, successful and above all, sophisticated – the sort of person you'd be delighted to have sitting next to you at dinner. Frankly, not everyone will get it (in any sense). Are we taking this 'coffee beyond compare' thing too far? We think not. [Name] is confident but never arrogant, cultured but never conceited, exclusive but never elitist. Just like you, really.

The tone of voice comes straight from the CEO and reflects the qualities of the product – defiantly upmarket but with an understated intelligence and sense of humour. It seems to work so I push on:

Secret No. 2 – Accept no limitation

It's a curious fact that many premium espressos don't work quite so well when combined with milk. Consider the fact that around 95% of espresso drunk in the UK ends up in either a latte or a cappuccino, and you'll see why that's something of a problem. [Name] coffee is different. It's a superb all-rounder with a taste profile deliberately created to produce a fruity, citrus espresso and a rich, chocolatey latte/cappuccino. 'So what?' we hear you cry. Well, no other coffee known to man is so versatile – you can under express it, over express it, drink it straight or with milk and it will still taste divine. Forget nanotechnology, that's what we call progress.

Again I'm rifling through my research to come up with mini-factoids like the thing about percentage of espresso drunk with milk and the taste profile (lifted directly from the taster's notes).

I'm on my way with the body copy but there's something missing. As it stands it's still all a bit, well, ordinary. What I need is to give the reader some reason to keep reading. Plus, I haven't really nailed the idea of 'premium' that comes up again and again in the research. How can I combine the two? Years ago I did a spoof questionnaire for another client which seemed to work, so I dig it out and try to graft that format onto this subject. And then it hits me – a multiple-choice questionnaire to find out if the reader is a 'premium person'. I go for the dependable 'one wrong, one right, one absurd' formulae:

Are you a premium person?

Find out with our handy test. Answers on the last page.

An awfully nice couple have moved in over the road and invite you round for supper. Do you take:

A. *A two-litre bottle of strong cider? Well, you want to break the ice, don't you?*

B. *A tin of [name]? Smart, subtle and exclusive, just like you.*

C. *News that their property is built over an old graveyard. Just because you often hear screams at night doesn't mean they necessarily will.*

Once I've got my eye in, the next instalment comes easily:

Your partner is of the opinion that premium coffee is a Barista-led conspiracy intended only to fleece the masses of their hard-earned crust. Do you:

A. *Agree before quietly dumping them at the first opportunity?*
B. *Patiently explain that although we are all in the gutter, some of us are looking at the stars?*
C. *Buy them a subscription to* Heat *magazine and tell them to lighten up?*

And so on. I write up all five 'secrets', come up with five quiz pages and then alternate them – one page of secret, one page of quiz. That's the first draft of the body copy done.

Now I need an intro and an outro. The intro needs to tee up the whole secrets thing, introduce as many of the key words as possible without seeming forced and generally set the tone. For some reason Stephen Fry starts talking in my head and I quickly arrive at:

Imagine a coffee that's smoother than a tiger in a tuxedo and more luxurious than a cashmere codpiece.

Are you imagining?

Well, that's [name], a remarkable espresso that is, quite simply, like no other. Given its unique character we might be excused for keeping the secret of our success, well, secret. *But we can't. We want all who share our love of fine things and even finer coffee to enjoy the singular experience that is [name]. So that's what this little book is all about – the inside story of [name], a premium coffee for, well, premium people. Enjoy.*

I'm on to something with the cashmere codpiece malarkey, so I jot down a few more:

More pleasurable than a mink glove massage
Sexier than a supermodel in a string bikini
Smoother than a well-buffed baby's bottom
As glamorous as a bling-encrusted ball gown
As indulgent as a spoon-fed chocolate soufflé
More cultured than a small army of art critics
More complex than a roomful of Rubik's Cubes
As versatile as a Swiss Army knife with added kitchen sink
As passionate as a frisky flamenco fanatic

Hmm, none are quite as good as the original, but I'll hang onto them just in case. As for the outro, well, that's the quiz results:

Are you a premium person?

You've finished our quiz, now learn your fate:
Mostly As: *You haven't quite got to grips with this good life thing, have you?*
Mostly Bs: *Splendid. You'll do nicely.*
Mostly Cs: *Ever considered minicabbing as a career? Maybe you should.*

Top and tail with a couple of quotes gleaned from the Web . . .

No one can understand the truth until he drinks of coffee's frothy goodness.

Sheik Abd-al-Kadir

A fig for partridges and quails,
ye dainties I know nothing of ye;
But on the highest mount in Wales
Would choose in peace to drink my coffee.

Jonathan Swift

. . . and it's in the bag. Time for a coffee.

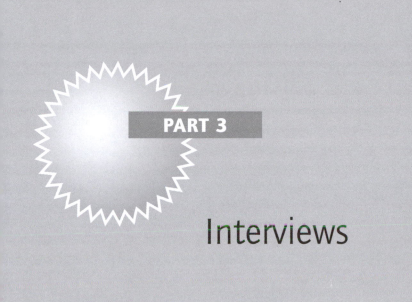

PART 3

Interviews

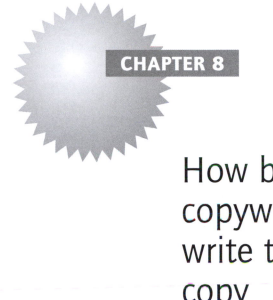

CHAPTER 8

How brilliant copywriters write their copy

I f you want to know about brilliant copywriting then it seems like a good idea to ask some brilliant copywriters how they work. So that's exactly what I've done. These interviews give a warts an' all insight into how copywriters really go about their business. I didn't want a polished, fairytale version of life as a copywriter; I wanted the grubby, dirt under the fingernails truth, and that's exactly what I got.

The people I've interviewed represent a broad spectrum of copywriting. Clearly there is great diversity, but one thing come across loud and clear: there's no arcane knowledge known only to initiates. Brilliant copywriting is about improvement by inches, not great leaps forward. It's about the patient accretion of ability through a process of practice and study. If that sounds suspiciously like hard work then don't worry; what also comes across is just how enjoyable this job can be. As a copywriter you've the opportunity to do incredibly satisfying work, mix with interesting, creative people and – if you're any good – earn a very comfortable living. They sound like good reasons to keep plugging away. So while there's no big secret to brilliant copywriting, there *are* lots of little secrets, as these interviews make abundantly clear.

At the end of each interrogation I've summarised what I think are key pieces of advice, and I've listed my faves below. It's no exaggeration to say that this material represents the distilled wisdom of some of the sharpest, smartest writers around. You may disagree with the particular points I've picked out as important – in fact I encourage you to challenge my choices and find your own favourites. What I hope is beyond dispute is the value and interest of listening to these copywriters talking candidly about their craft, their ideas and their ideals.

Common themes

Naming names

There's no overwhelming agreement on what to call this thing we do. It's true that 'copywriter' comes out top, but there are plenty of alternatives. I must admit I'm not sure myself. Odd, isn't it? Most professions know exactly what they're called.

Getting started

It's the same story over and over again – people fall into this job rather than deliberately seeking it out. Most interviewees profess an early love of words and a desire to earn their crust that way, but many also said they didn't realise that was a possibility. Perhaps we need to work harder spreading the word about our profession to young people instead of leaving them to blunder into it unaided.

Early influences

Many interviewees mentioned a significant boss who helped shape their thinking. If you've got such an individual in your organisation I suggest you seek them out and beg/blackmail them into acting as some sort of mentor figure.

Another theme is the way many of these writers managed to turn their early experience – often only tangentially connected with copywriting – into the foundation of their professional approach.

Personality traits

Most interviewees read voraciously, and they're endlessly curious about the world. Beyond that it's pretty much up for grabs, although a couple of people did emphasise that copywriters tend to be introverted curmudgeons. Guilty as charged.

Influences

Again, many opinions, little consensus. The admen David Ogilvy, David Abbott and Tim Delany come up a couple of times, as do writers Graham Greene, P.G. Wodehouse and George Orwell.

Planning and preparation

The general approach seems to be to read everything and more, then let the magic happen. A little more helpfully, many interviewees talked about not starting until they're full to bursting with ideas and know exactly what question they're trying to answer. Formal techniques like mindmapping don't figure highly. Copywriting really seems to be an instinct-led activity, which slightly undermines the aim of this book, but there it is.

Tips and tricks

My favourite has to be the Churchill quote ('Begin strongly, have one theme . . .') provided by Will Awdry. I'm also very taken with yoga lady Sarah McCartney's suggestion that standing on your head makes the ideas fall out.

Creative thievery

It seems we all do it, and hurrah for that. The collection and recycling of choice phrases, ideas and approaches seems to be part and parcel of our work. I can't improve on Jean-Luc Godard's maxim, 'It's not where you take things from, it's where you take them to.'

Creative block

The main suggestion seems to be 'talk to someone about what you're trying to do' – it'll force you to restate the brief, and in doing so you might find another angle. Engaging the other side of your brain with some poetry, web browsing, walking around the block or staring out of the window also figured highly.

Advice on getting into copywriting and improving your writing

Just do it. The theme of 'practice, practice, practice' came up several times. Copywriting is definitely a craft you learn by doing, although 'doing' can also mean reading the best books on the subject, talking to the best people and generally getting involved in the whole world of creative communications.

Top tips from top copywriters

No list of pithy proverbs is going to make you a brilliant copywriter, except perhaps this one:

- As copywriters we're basically in the meme business.
- Be brief; the brain is a cognitive miser. It's too busy thinking about love or lunch.
- Churchill said it all: begin strongly, have one theme, use simple language, leave a picture in the listener's mind, and end dramatically.
- Don't start writing until you're genuinely excited by your research.
- Get used to having your work mangled.
- Go as far as you can creatively, for your own sanity as much as anything. The client will always bring you down to earth.
- Great designers are often great writers. Learn from them, particularly the way they control the creative process.

- If it takes too long to write then it's probably wrong.
- If you're stuck, explain what you're trying to do to someone unconnected to the project. That'll force you to come at it from a different angle.
- Like poetry, copywriting is about using as few words as possible to say as much as possible.
- Never underestimate your readers' intelligence.
- Pick your battles. Not everything is worth fighting for.
- The backstory isn't superfluous; often that's where the real power lies.
- The best copywriters tend to write like they speak, or rather how they'd *like* to speak.
- The brand is the star, not you.
- The first sentence is the most important, but the second has to work even harder.
- The more senior the people you work with, the faster you'll progress.
- The reader is everything. Agree who they are early on.
- The real work is getting the thinking right; the rest is colouring in.
- Think about who's ultimately signing your work off – what do they want?
- To become a better writer, you need to become a better reader.
- What's the story you're trying to tell? If you don't know, don't start writing.
- Wit works wonders, just don't overdo it.
- Words are ideas, and ideas are the key to brilliant copy. Get plenty of ideas down, then edit with cruel brutality.
- Your favourite phrase probably has to go.

Interviews

⟩ **brilliant** questions and answers

Tim Rich

'Copywriters work in this incredibly fertile, endlessly fascinating area between their clients and their readers. They act as translators. So brilliant copywriting is the compelling, believable expression of what the company has to say in a way the reader wants to read.'

To east London, and Tim Rich's live/work weavers' cottage at the very end of Brick Lane. Built in the closing years of the eighteenth century, the whole row were originally connected by now bricked-up internal doors, effectively turning the terrace into a cloth production line 100 years before Ford's automotive adventures. But I digress. Tim's focus on corporate communications provided an unusual perspective on the craft of copywriting. Not for him the fast turnaround, marketing-orientated work that fills so much of my time. Tim's work is deeper, slower and more detailed, which brings its own set of challenges. Here's what he had to say.

Q What exactly do you do and how do you describe it?

A I'm a writer and editor for businesses. I specialise in corporate communications, so I write speeches, annual reports, strategy papers, corporate tone of voice stuff and everything that flows from that. My work is about how the business communicates with the world, rather than promoting specific products and services. My audience can range from a single environmental protester to a huge pension fund.

Q What are the main differences between your work and what's sometimes called marketing copywriting?

A It's a longer form. The sustained argument is the bedrock of corporate copywriting. Plus, I usually work with senior people in the corporations I deal with. They take writing very seriously because they're aware of the influence their words can have.

Q How did you get into copywriting?

A I was a design journalist so I met lots of great designers – people like Alan Fletcher and Mike Dempsey – who had a sophisticated idea of what copywriting should be doing. It was an extraordinary blast of influence. So when I got into copywriting the people I was learning from were designers who had a real feel for words and were often doing great writing themselves.

Q Was there anything you learned on early jobs that you've used ever since?

A Certainly how to cope with scale – I learned to manage the sheer volume of information and build my argument. Another thing I learned is that there's no room for being temperamental. The focus of everything must be the reader. And although I'm clearly on the side of the client – not least because they pay me to represent their interests – they don't pay me to be one of them. They employ me to sit between them and their readers. It's tempting for copywriters – all copywriters – to go native and start using the phraseology of the clients. If you go too far that way your value to your clients – and your reader – rapidly diminishes.

Q You mentioned how important words are to corporations, which obviously puts you in a sensitive position. How do you establish trust with a new client?

A It comes down to listening. I don't mean to sound like an American greeting card, but listening is everything and I'm constantly struck by how often copywriters just don't seem to really listen into the words clients say in briefings and interviews. Listening comes in two forms – what they said, and what they meant. You've got to get both.

Q Are there any personality traits that characterise brilliant copywriters?

A Most copywriters I know are introverted, and that can be a problem. Business is increasingly presentational, so the writer really needs to be able to stand up and talk about his or her work in an intelligent, articulate way – and invite criticism and comment. That can be a challenge.

Q You've worked as both a journalist and copywriter – what are the big differences?

A In an article you're presenting certain points of view that build towards a particular conclusion, all clearly authored by you. A piece of copy for a client has to leave the reader with a clear and consistent set of points – it's about what the client needs to get across, not what you want to write about or what the pros and cons are. The value of copywriting lies in what lodges in a reader's brain and what they take away with them, so it is very much about defining and expressing memorable messages.

Q Based on your experience as both a journalist and a corporate writer, have you evolved any tips that could help junior writers?

A You need to get everyone who's influential on the project to agree who the readers are. It's amazing how often that doesn't happen. Reading your words out loud is also enormously helpful – if it's not easy to speak it won't be easy to read. Also, don't be afraid to ask stupid questions and to be the only person in the room who disagrees.

Q How do you get things moving when the words just won't come?

A If something isn't working then I know myself well enough to not try and force it through. I'm far better off taking a half-hour break and going for a walk or reading a book. But I also talk to people. Usually I'm stuck because I'm missing vital information or I've misunderstood something or I've got sucked into some kind of limitation. So phoning the client can be amazingly effective at clearing block. It's about getting another point of view to shake stuff up.

Q How do you plan and prepare for a new job?

A Just read everything about the company you can get hold of. Here's a tip: if you want to know about a company quickly go to the press release section of their website. Press briefings are written for journalists, many of whom are considered hard of understanding, so very quickly you can find out about what's going on. I hate starting until I've got the whole argument mapped out and agreed with the client. I'm trying to bring the argument to life, and if I'm not clear of the argument I shouldn't be writing. So I write to a plan that defines what every paragraph should say.

The best of these are so detailed the job could be completed by another
writer if I were to be knocked down by a bus.

(Q) How do you know when something is working?

(A) Instinct. If you're taking ages to write a sentence or paragraph there's
something wrong. You need to stop and talk to someone or read poetry or
scream or whatever. And that's about trusting your instincts. When I started
out I spent too many hours trying to claw my way through texts. I'd have
been far better having the confidence to just stop and think for a while.

(Q) Do you actually *enjoy* writing?

(A) I love the feeling of writing a narrative where the argument is just
flowing and I can hear the chief exec saying it. Day to day the bit I love
most is editing. I think brilliant copywriters are often brilliant editors. The
words that aren't building the argument are the enemies of the other
words. They're poisoning the whole piece, so they've got to go.

**(Q) And finally, imagine you were mentoring a younger writer. What
advice would you give them?**

(A) Most of what makes a good copywriter is the effort they're prepared to
make. Readings, professional courses, talks – they're not a 'nice to have',
they're essential for professional development. Also, don't become a
freelancer too early because there are tremendous advantages to working
for a company, getting training, learning the business and making contacts.
And if you do go freelance, don't neglect the real benefits of your job –
things like the flexibility to travel, study or write other stuff.

In a nutshell:

- Great designers are often great writers. Learn from them,
  particularly the way they control the creative process.

- Don't act like a highly-strung prima donna. If that's really
  you then you're in the wrong job.

- The reader is everything. Agree who they are early on.

- Maintain a slight distance from your clients. It aids objectivity.

- Learn to listen.

- Think about the context in which your words will finally appear and let that influence the writing.

- If you're stuck try talking to the client.

- Don't be afraid to be the only dissenting voice. Just be good-humoured about it.

- Press releases are a fast way to get inside a business.

- If it takes too long to write then it's probably wrong.

- Copywriting is all about making an effort.

↗ brilliant questions and answers

Will Awdry

'Brilliant copywriting conveys an idea or thought so extraordinarily memorable that once I've read it I can never forget. It engages me with a subject I had no knowledge of or interest in and would otherwise have walked past. It just stops me in my tracks.'

I shared the lift up to Ogilvy's Docklands HQ with a young photographer clutching a portfolio the size of Rutland. We got chatting about what I was there for. 'You'll get a good interview,' he said, 'Will's the nicest man in advertising.' And so it proved. After a brief chat, Will – who is indeed a distant relation of the Rev W.G. Awdry of 'Thomas the Tank Engine' fame – made me a cuppa and led me to his office (that's how important he is). After the obligatory shuffling of chairs and switching on of tape recorders we got down to business.

Q **Perhaps you could start by telling me what exactly you do.**

A I'm the joint Creative Director at Ogilvy in London. I've spent most of my working life in advertising, initially as an account exec but for the last 23 years as a copywriter and creative director. These days I'd have to call myself a creative director, although I'm still a copywriter in the sense that I'm involved with selling, mainly selling ideas to my own people – the most sceptical audience of all.

Q Do you miss hands-on writing?

A Yes, but there are pros and cons to both positions. Advertising copywriting places a real premium on the idea, so these days I'm coaching people whose copywriting has to build those ideas in the mind of the audience. Plus, there's no point having a dog and barking yourself. You need to establish very clear rules when you're guiding and managing creative people. If I half write something, either in conversation or in front of their eyes, it's important they can then take ownership of it. That generosity of spirit is essential and works both ways.

Q When you started out was there anyone there who took you under their wing?

A Barbara Nokes was Head of Copy at BBH where I began and every single thing you wrote had to go in front of headmistress. In that situation it was better to be an empty vessel and just absorb everything around you. I also voraciously read anything to do with creative advertising and commercial communications. For about two years of my life I could tell you who won Best Art Director in Ulan Bator in 1957 for 'Best advert for household appliance made out of yak milk'. I became deeply dull.

Q Was seeing your first ads in print a thrill?

A Yes, it was an extraordinary feeling. It's a thrill that diminishes a little as time goes on, I must say. There's still a faint echo of that today, but what's really going to make you famous isn't your name in lights, it's getting other people's names in lights.

Q In what sense?

A If you look at the work of David Abbott, a hugely celebrated copywriter, he had a clear house style. But what you remembered in his work for Sainsbury's or Volvo or *The Economist* is the distinct language he created for each. If you really wanted to see your name up there you'd be better off writing fiction.

▶

Ⓠ **So it's about making the brand the star?**

Ⓐ Exactly. If you make the brand the star, you're doing well at your job, and one piece of advice I give to my students is 'I don't want to see one piece of your writing because that's you talking to me. I want to see at least ten products/brands/services or organisations talking to me to demonstrate you can become different voices.'

Ⓠ **How do you actually start writing?**

Ⓐ There comes a point where all the cups are washed, the desk is tidy, all my pencils are sharp and there's nothing I can do to delay the moment any longer. It's about panic and a naked fear of having been seen to have failed. Once I've got something down on paper I'll leave it alone. After minutes, hours or days I'll reread it and think, 'Hmm, that's OK but that needs to go and the order needs to change and I need more here' and so on. Almost certainly the bit you're most in love with – that choice turn of phrase you're so proud of – has to go.

Ⓠ **Any tips or tricks you'd care to share?**

Ⓐ I've lots of little tricks that are begged, borrowed and stolen from other writers, and that's the generosity of spirit I was talking about. The best advice I know comes from Churchill: begin strongly, have one theme, use simple language, leave a picture in the listener's mind, end dramatically.

Ⓠ **That's my whole book in 17 words. What about big influences on your writing?**

Ⓐ Certainly Barbara Nokes who guided me in a Miss Jean Brody style during my early years. I'd also be mad not to acknowledge people like David Abbott and Tim Delany. Beyond advertising I'd say Richard Ford – who was David Abbott's favourite writer – and Graham Greene – who was Tim Delany's favourite writer. It's his economy and the amount of mood he creates given the lack of adjectives that make Greene's writing so extraordinary.

Ⓠ **How do you stay fresh?**

Ⓐ I'm always looking for new choice phrases, and like a magpie I nick stuff all the time. The writer that's influencing me most right now is Marina

Hyde in the *Guardian*. Her use of language is fantastic – she has fun with words. I think reading is important for your sanity and makes writing more fun.

Q How do you feel about being a bit of a magpie? I think we all do it.

A This touches on an oft-visited but seldom resolved subject – and that's what's yours and what you contributed. We all have the same narrow range of tools at our disposal – 26 letters – and we inevitably borrow and are influenced by others. It's a question of use and how we recontextualise what we pick up. You can't put a Berlin Wall around your thoughts.

Q So is originality important? Or is authenticity more significant?

A Originality is still possible, even in the most tired, overworked areas like car advertising. God knows how much time, money and effort have gone into selling cars over the years, and yet I still see fresh thoughts and inventive new ways of coming at the old problem of flogging cars. So originality is never dead. Having said that, one person's originality is another person's reworking of existing thought, that then comes across as something emphatically authentic. That gives it a definitive quality and the appearance of originality.

Q How do you know when you're ready to write?

A The first thing is, if you don't know what question you're trying to answer you shouldn't start. So ask yourself, 'What am I trying to do?' Once you've got that you can formulate your answer. And if you know where you need to get to, it doesn't matter where you start. Maybe you start on what ultimately becomes paragraph three, maybe somewhere else – it doesn't matter. But you need to know what you want to achieve.

Q How do you organise longer pieces of writing?

A I might do a flow diagram to work out what I need to establish before I can make my killer point. I might then end with a flourish that reiterates the introduction – the so-called well-made ad. I do tend to start at the beginning and write through. The real test is how much can I take out before it falls apart, because that's when you realise how much of your own personality you've put between the message and the reader.

Q **Do you ever get stuck? And if so how do you break out?**

A It involves a third party. Find someone unconnected with what you're having trouble with. Don't explain the problem, but instead explain what you're trying to do, the message you're trying to put across. You'll usually have to go back to square one to explain it, and in that conversation you'll come at it from a different angle.

Q **How do you know when something is working?**

A You get to a point when it's almost impossible to change something because everything is important and everything matters. You go for a walk around the metaphorical block, come back, look at it and think, 'I can't really do any better than that'. In fact that probably means it's overworked already. As you said before we started, first drafts tend to have an immediate energy that you've got to be careful not to lose.

Q **As a creative director you're involved with hiring and firing. Any advice for a copywriter trying to break into or progress in advertising?**

A Your portfolio is your calling card, so you need to show a diversity of voices and approaches. Start by approaching as many people as you can, then narrow your mentors to just two or three people as soon as possible. There are as many opinions as there are individuals and you risk being paralysed by contradictory advice if you're not careful. Also, work on your technique because there's only so far you can go on instinct. Being grounded in the metre of what you're trying to get across will help. Copywriting is a craft skill that needs regular transfusions of originality from the real world.

In a nutshell:

- Copywriting is the craft of placing ideas in the mind of the audience.
- Voraciously read everything and anything to do with creative advertising and commercial communications.
- The brand is the star, not you.

- If you're stuck, explain what you're trying to do to someone unconnected to the project. That'll force you to come at it from a different angle.
- Get used to having your work mangled.
- Your favourite phrase probably has to go.
- Churchill said it all: begin strongly, have one theme, use simple language, leave a picture in the listener's mind, end dramatically.

↗ brilliant questions and answers

Nick Asbury

'Brilliant copywriting never feels overly written. It persuades you without being obviously persuasive. That usually means telling people what you want them to know in the simplest, most human way possible.'

To Manchester's achingly hip Northern Quarter to meet Nick Asbury, a marketing copywriter with a broad range of clients and a furious turnover of jobs. He's also the creator of some superb pieces of promotional material including Pentone, a cool spoof of the Pantone chip idea using different tones of voice instead of colours, and Corpoetics, a collection of 'found' poetry from the websites of well-known corporations. He's written a book – *Alas! Smith and Milton* – about the agency of the same name and runs Asbury & Asbury, a creative partnership with his wife, designer Sue Asbury. What a busy chap.

Q How do you describe what you do?

A When you say 'copywriter' people assume you work in advertising, so I usually say I'm a copywriter for design. A good 80 per cent of my work comes from design companies and branding consultancies. So the designers do the visual side of things and I do the verbal side of things. You could call it branding with words rather than images.

Q How did you get started?

A I did an English degree and finished that without a clue what to do.

Eventually I saw an ad in the *Guardian* for a graduate trainee copywriter at a recruitment advertising agency, a very unglamorous corner of copywriting. I got the job and spent six months writing recruitment ads. I remember rushing out to buy the paper to see my work. I think my mum and dad pretended to be more impressed than they were.

Q Did you learn anything really important doing those early ads?

A I think doing recruitment ads is good training for copywriters. You have a very tight brief – who, what, where and so on. So the first things I learned were about structure and reducing things down to fit a quarter-page space or whatever. It teaches you discipline and how to prioritise messages.

Q It's sometimes said copywriters are frustrated novelists. Sounds familiar?

A Not exactly. My main interest outside copywriting is poetry. I think having interests beyond copywriting definitely gives you an extra set of reference points. There are lots of parallels between copywriting and poetry, not least because poetry is about using as few words as possible to say as much as possible, exactly the same as copywriting. Plus, it's a great source of phrases to steal.

Q Tell me about your Corpoetics piece

A I was going through a phase of reading lots of poetry and books about poetry, and I came across a reference to found poetry. I thought, 'that's interesting', and somewhere in my brain something clicked and I realised I could make found poetry out of corporate copy. So I quickly went to the About Us section of McKinsey's website – the most corporate thing I could think of – and set myself the task of using only those words. Then I did it again and again with other companies' copy. I did it for pleasure but it worked really well as a promotional piece, which I sort of suspected it would all along.

Q And the Pentone thing was in a similar vein ...

A Yes. It's not a case of thinking 'I need to do a business mailer, how can I be quirky?' but rather I'm constantly surrounded by design paraphernalia and at some point the penny dropped. I realised I could change 'Pantone' to 'Pentone' and do something on tone of voice. I have to say that with both Corpoetics and Pentone it really helps that my wife is a designer. In fact, marrying a graphic designer is a great career move for any copywriter – she does my website, mailers and so on ...

Q My next question is about any tips you've got, and clearly the first is 'marry a graphic designer'. Any others?

A One of the best I've got isn't so much a writing rule as a general business rule, and that's 'pick your battles'. You're always engaged in this struggle between trying to do brilliant, creative work that can go in your portfolio, and doing whatever will make the client happy so you can get your invoice out. I remember being told early in my career that every job is a potential award-winner and that you should fight for it with the client. And I gradually came to realise that's not good advice, because if you really follow it then you'll drive yourself mad. I freely admit that some jobs are fundamentally dull and just need doing as well as possible and showing the door.

Q Who's influenced your thinking as a copywriter?

A I came of age professionally at a time when wit was paramount, and the book that captures that is *A Smile in the Mind* by Beryl McAlhone and David Stuart. On the other hand I do think you can become too obsessed with witty ideas. Sometimes the answer is just to do a well-crafted piece of copy that conveys the right messages and embodies the right personality.

Q What's the hardest part of a job?

A Getting the first sentence or paragraph right. I might rewrite it 30 times. That's why I use a computer so I can edit easily – if I worked with paper I'd get through a small wood every day. Actually, I'm now going to contradict myself, because the hardest part of being a copywriter is all the stuff that goes around the writing. As well as being a copywriter, I'm effectively the account manager, new business person, finance department and so on. That's what keeps me awake at night, not the writing.

Q Every job starts with a brief of some sort. What separates the good from the bad and the ugly?

A I don't mind too much whether the brief is narrow or wide open – what matters is that it's clear. It makes writing so much easier. With 70 per cent of jobs you don't get that, so the first stage of your job is to clarify the brief. Most of the briefs you'll receive over your career won't be very good – that doesn't mean you should tear them up and walk away; instead you need to develop a knack of seeing *through* the brief to what they really want to say. My grandad used to tell me about his army training exercises, where they'd be walking through fields looking out for snipers in the undergrowth. His sergeant would tell him 'Don't look *at* the bushes, look *through* the bushes.' It's the same principle.

Q How do you plan and prepare?

A I record a lot of meetings and transcribe the important bits. I then scribble down notes and draw circles around the best bits. So the argument almost forms itself out of a cloud of chaos. The thing is, for some jobs I almost feel I could write the piece after one phone call and that the extra meetings are more for their sake to offer reassurance. That said, you have to guard against complacency and formulas. You need to keep it fresh somehow, as much for your own interest as anything else.

Q And on that subject, how do you break out if you're stuck?

A If I'm stuck it's usually because I'm uninspired rather than unable to crack a tricky creative problem. So I leave it and go for a walk and come back to it when I'm feeling more upbeat. It's the exact same job but somehow it's become more inspiring in the meantime. I do try to write when I'm in a good mood; if I'm feeling turgid I'll try to do invoicing or something. Or I might break off to read a poem, or even write a poem, just to engage the other side of my brain.

Q How do you know when something is really working?

A It's when you know what the next sentence is going to be before you get there. About halfway through a good piece, I'll also realise what the last sentence is going to be, so it becomes a pincer movement. You just need to fill in the rest then. That's not to say there isn't a lot of humdrum stuff along the way where it feels like walking through treacle – you definitely shouldn't become a copywriter if you're not prepared to do the boring stuff.

(Q) Any sage advice for young writers?

(A) Just start. Take some press ads and rewrite them, start a blog or website, write articles for a magazine – anything to help you stand out. Or just write a really good letter to the top design and writing companies and ask if they want an in-house junior. You should also join 26 and get to know people. But mainly just get writing.

In a nutshell:

- Copywriting is branding with words rather than images.
- Having an interest in language beyond copywriting can give you an extra set of references.
- Like poetry, copywriting is about using as few words as possible to say as much as possible.
- Pick your battles. Not everything is worth fighting for.
- Don't look at the brief, look through the brief.
- Think about who's ultimately signing your work off – what do they want?
- Wit works wonders, just don't overdo it.
- Marry a graphic designer.

↗ brilliant questions and answers

John Simmons

'Brilliant copywriting creates emotion in me as I'm reading – a smile, a lump in the throat – anything. It's also about the pleasure of something that's just right – all the words are in the perfect place and you wouldn't want to change one bit.'

If there's one person responsible for raising the profile of business copywriting and writing for design over the last decade it's John Simmons. He might blush, but the influence of his books and the *Dark Angels* writer training courses he helps run can hardly be overstated. He's also one of those remarkable people who seem to know everyone. Forget six degrees of separation, John is more a one degree of separation chap. Always absurdly encouraging and generous with his time, I met John in the café at the Royal Society of Arts in Covent Garden where we chatted over the clink of cups and saucers.

Q I know you describe yourself as a business writer. What's the difference between that and a copywriter?

A Not much. Your description of copywriting as 'writing that sells' is a good starting point, although it could be argued that most writing for business is also to do with selling something, even if it doesn't involve the exchange of money. I think business writing is broader, while copywriting is more specialist. Business writing also tends to be longer in terms of word count. Even long copy ads are actually quite short compared to a lot of what I do.

Q On that subject, how much writing do you do these days?

A Quite a lot. I think it's important to keep my hand in and it helps me to write the books – after all, who am I to speak if I don't do it myself? I do a lot of work with design agencies, mainly with people from my past. Since I left Interbrand in 2002 – I was director of verbal branding there for several years – the wonderful thing that's happened is that it's freed me to work again with a lot of people I knew in my earlier days, like Domenic Lippa and Harry Pearce who are now partners in Pentagram.

Q You've clearly resisted the tendency among senior people to step back from hands-on work and concentrate on consultancy . . .

A Yes, I have resisted, mainly because doing this kind of writing keeps me in touch. It keeps me sharp. And things develop – the copywriting of today is different from the writing of ten years ago, so you have to stay in contact with that. But I'd also say that a lot of the writing I do is a form of consultancy – it's writing to think through a problem and come up with the best words to solve it.

Q How did you get started?

A At university I wanted to be a novelist. That didn't turn out to be a practical career so I got this strange job working for the National Economic Development Office where I was communications manager. They produced reports about the clothing industry or the retail industry or whatever. They were pretty dull, so I had to make them accessible to the people working in these businesses. What I really learned was how to turn industrial gobbledygook into language that ordinary people could understand.

Q I've often thought that much of the writing we do is really translation . . .

A I absolutely agree. We're translating from the sort of words understood by one group into the sorts of words that would be understood by a much wider circle.

Q What sort of hours do you work?

A Years ago one particular firm of designers called Lock Petterson asked me to do some moonlighting from my day job. And that's where I got into this habit of working that I still maintain. I had a full-time job and two young children who I wanted to give as much time to as possible, so I had evenings and weekends free. I would get home from work, put the kids to bed and then work from 8 p.m. on Friday evening to 2 a.m. on Saturday morning – six hours of concentrated writing. That worked for me and I've continued to do that ever since – it's my time to do my most important writing.

(Q) **Let's change tack slightly. Any personality traits that mark out copywriters?**

(A) I've thought about this before but I've not come to any firm conclusions. It's not even an obsession with the craft of writing because I've come across writers who don't share that. Copywriters can be very different people, and I think copywriting needs that. Actually I'm going to contradict myself now and say there's an element of being comfortable with being a bit of a loner, but then lots of copywriters work well with art directors or designers so even that's not always the case.

(Q) **And on that subject, do you have any golden rules for better writing?**

(A) I think there are some fundamentals – often fairly obvious ones. For example, write with personality, draw on your own experience and put it into your writing. I get quite shamelessly emotional in my writing at times. In fact some of the best writing I do with tears in my eyes, not just brought on by the fact that it's 1 a.m. and I want to get to bed.

(Q) **Touching people like that is unusual in copywriting ...**

(A) Well, while it's important to be objective in some forms of copywriting, some of the best writing works by being emotional rather than purely rational. You need a good rational case whether you're writing an ad or whatever – David Ogilvy talked about the importance of doing research and getting your facts straight, but then you have to bring something else to it. If that wasn't true then PowerPoint slides full of bullets would actually work.

(Q) **Talking of Ogilvy, who are the key influences on your writing?**

(A) It starts with my school and university days. I'm a writer now because of the things I learned then, so my influences are from the world of literature, particularly Dickens and Shakespeare. Amongst modern writers I like John Irving, F. Scott Fitzgerald, Patrick White – an Australian writer who won the Nobel Prize in the early seventies. For me to become a better writer, I need to become a better reader, so I've always a book on the go. When I first got serious about copywriting I read Wally Olins' *The Corporate Personality* and Ogilvy's various books.

Q What tools do you use?

A I always do my first drafts in longhand using a pencil. On paper or in a Moleskine notebook. Later I write everything up on a computer, but writing longhand gives me a closer connection with my thoughts. You were asking about golden rules, well that's a golden rule for me.

Q What's the hardest part of writing for you?

A Working on anything that doesn't have any meaning for my life. I've never got on with working for financial companies, although I can't avoid them.

Q Short copy or long copy?

A I like longer pieces. I like what you can develop in a longer piece in terms of logic and the ability to really influence people emotionally – I find it very hard to do that with a strapline. Even the best examples you can think of, you can imagine a backstory behind them that would make them much more powerful. For example, 'Just do it' doesn't mean much, whereas if I wrote the story of Roger Horberry going to Amsterdam and running the marathon, and the achievement and commitment behind that, there'd be much more impact.

Q What sort of research do you do?

A Reading, mainly. I obsessively note take. I then read through my notes highlighting what I think is really important and use that to structure my argument. So I plan in my head and write my first draft directly out of my rough notes. I said reading is important, but talking to people is even more essential. They say really significant things without realising it – it might just be a word or a phrase – and suddenly there's your answer.

Q Where do you start when you're writing?

A I start anywhere. Beginnings are important, but often you find your idea for a beginning half-way through or at the end. Last week I was writing a brochure on personal finance for Pentagram and I'd spoken to someone who made this analogy about shoes. So I started with the idea of shoes just to see if I could do anything interesting with it. It might turn out to be a small part of the finished thing, but it's a way in.

Q **If you're stuck, how do you break out?**

A The best way is just to start. Spend ten minutes just writing whatever comes into your head and go into a Zen-like state as you do that. After ten minutes look at what you've got and you might have a load of rubbish or you might have something really worthwhile. I've a lot of books on the shelves at home so perhaps they help. For example, if I'm writing about a spa I might open up a book like Patrick Suskind's *Perfume*, only to discover it's a great book about how to be a psychopathic murderer rather than an investigation of the senses.

Q **How do you know when something is working?**

A Sometimes I go through a stage that's quite close to anger. I've got all these resources and notes and I've worked myself almost into a state of rage because it won't come. And then something suddenly emerges from this and the anger subsides and I know I can get on and do it, and that's when it's happening.

Q **What's the best way for someone to improve their copywriting?**

A Practice, practice, practice. The more you do it the better you get. Reading books about copywriting is important. Reading books that *aren't* about copywriting is even more important. And you've got to want to be a writer – actually, going back to one of your earlier questions perhaps that's the only thing all copywriters have in common.

In a nutshell:

- Copywriting *sells*, even if that doesn't involve the exchange of money.
- It's often about translating from one vocabulary into another.
- Don't neglect the emotional when constructing your argument.
- To become a better writer, you need to become a better reader.

- The backstory is far from superfluous; often that's where the real power lies.

- An unrelated word or idea can provide a brilliant way in to a brief.

- A few minutes non-stop writing can break block.

↗)brilliant questions and answers

Sarah McCartney

'Brilliant copywriting makes you want to read to the end. No matter what it's about, you start at the beginning and you think 'ooh' and you just can't stop.'

Walk down the high streets of Britain and sooner or later you'll be assaulted by a rich cacophony of fragrance emitting from a vivid yellow shop front labelled 'Lush'. Actually 'fragrance' is the wrong word – Lush deal in big, lusty smells so good you want to pop them in your mouth and suck slowly. Since 1996 Lush has promoted its products through the *Lush Times*, a periodic publication given away to customers and sent to a mailing list of around 200,000 cosmetics lovers the world over. The copywriter behind *Lush Times* is Sarah McCartney, an insanely hard-working writer and trainer. As luck would have it Sarah happened to be in York on business, so after the day's work was done she came round for a coffee.

Q Call yourself a copywriter?

A It depends on the day. I have two business cards. One says 'Copywriter', the other that I'm a 'Strategic Storyteller' because I like that and it interests people. If people ask, I mostly say that I write for a living and run writing workshops.

Q How did you get started in copywriting?

A I did science at university and then went into advertising on the statistics and analysis side. I switched over to marketing and worked in the *Guardian*'s marketing department. Around this time I came into contact with the guys from Lush after they made some products for a *Guardian* Valentine's Day promotion I was running. They asked 'Do you want to write

Lush Times?' So I said yes and from that day I was a writer. That was in 1996.

(Q) How much do you actually write these days?

(A) The *Lush Times* is 40,000–50,000 words and that has to come out four times a year. That's mostly me. I'm allowed to recycle some of the words because Lush is very environmentally friendly, so I don't have to write the thing entirely from scratch every time.

(Q) What can you remember about the first thing you wrote that got used/produced?

(A) I wrote a piece on how to pass your bike test for *Weekend Guardian*. I was told to write 1000 words and I got it down to 4000. It took me ages to get it finished. Anyway I wrote this piece and it was edited and appeared with a great big picture of me on my motorbike. At that point I was officially a writer. It was a milestone – actually, it was more like Stonehenge.

(Q) What did you learn from your early pieces?

(A) By the time I did the second one, when they said they needed 300 words I knew to give them 300 words: editing was definitely lesson number one.

(Q) Any golden rules you've arrived at over the years?

(A) If you're stuck go for a walk and go on the swings in the park. I do believe that standing on your head causes ideas to drop into it – a yoga thing. I'd like to run joint yoga and writing courses because I think people would be full of ideas afterwards. Another thing I like to do is open a book at random, stick my finger down the page and find a word and fit it into the next thing I have to write. It's not about creativity; it's about wiles; it's a technique.

(Q) Any big influences on your writing/thinking?

(A) David Brook, boss of the marketing department at the *Guardian* – he made us present all proposals to him on a single sheet of A4 with bullet points. Also Nancy Banks-Smith, one of the *Guardian* critics – I read her television reviews and thought, 'Oh my God, we're allowed to write like that'. She didn't approach it in a scholarly way: she approached it as a genuine human being. That's what I've always done.

Q **How about the whole short copy vs. long copy thing?**

A I'm very good at anything from paragraphs upwards. I believe that long copy is sadly neglected. People say that there's no point writing long copy because people don't read it, but that's a self-fulfilling prophecy.

Q **Do you ever get really stuck? And if so, how do you break out?**

A I was stuck yesterday. I get stuck when I'm very tired. I don't believe that people's best creative ideas come when they are exhausted. To get unstuck I think about what's happened today and start writing for three minutes: what did you notice today, what was unusual ... I just get it down. I'm better if I know somebody is waiting for me and I owe them it.

Q **Do you read much?**

A I love reading 1930s detective novels for the language, whether it's Damon Runyon and his American gangster language or Margery Allingham or Edmund Crispin. I go to them if I ever feel like I need topping up.

Q **How long does an edition of the *Lush Times* take to write?**

A It used to be two months full on, so around forty working days producing 1000 words a day. These days it's very unlikely that I get that much time because Lush is so much bigger now. So I end up with fewer days with more to pack in. It's more like twenty days of 2000 words each, and even that speeds up at the end.

Q **And finally, what advice do you have for anyone trying to become a copywriter?**

A Start now. Sell a few things on eBay. Write copy for those. Don't make it sound like an ad; make it sound genuine. Write the way you would tell it to somebody. Don't allow long words to sneak in and make it all writerly. Practice, practice and practice some more. Nobody is going to give you a job if you haven't written anything.

In a nutshell:

- If they ask for 300 words, don't give them 1000. It's not being generous, it's being wrong.
- If you're stuck, try picking a word at random and working that into whatever you're writing.
- You've got a choice: sit there feeling guilty about not doing it, or just get on with it.
- Write like you'd say it. The more writerly something is, the less human it'll feel.
- Very few people can create effectively when they're dead tired. The all-nighter is a myth; don't fall for it.
- Standing on your head causes ideas to drop into it. Possibly.

↗ brilliant questions and answers

Dan Germain

'Brilliant copywriting doesn't demand any explanation. It's short and sweet and hits the spot first time. Just recently we had a competition to find some words to go on the bottom of our bottles. The best was "Trapped in bottle factory. Send help." Now that's brilliant.'

What can we say about Innocent's much-praised writing that hasn't been said before? Hopefully something or you might as well skip to the next interview. The fact is their quirky text has put copywriting on the map for many people, although their success has caused copywriters some problems as these days every other client, regardless of their business, wants to sound like Innocent. To find out more I headed for unlovely Shepherds Bush to speak to Dan Germain, Head of Creative at Innocent. He met me in reception minus shoes and socks, presumably all the better to experience the tingle of Astroturf between his toes – you didn't imagine Innocent went for anything as conventional as carpet, did you? After a few stiff smoothies we got to work.

Q **What do you call yourself?**

A I'm Head of Creative at Innocent Drinks. I've worked here since it started in 1999. In that time I've helped take Innocent from something people knew nothing about to something that more people know about, and a lot of that's hopefully been to do with stuff we've written. I have other creative duties to do with design, but writing is the thing I'm proudest of and I think I'm best at.

Q **How much writing do you do these days?**

A Less than I used to. I have a team that includes a dedicated writer called Ceri, so a lot of what I do is rewriting, editing and coaching. I tend to still write bigger stuff like adverts. Plus at the moment we're writing a book on Innocent and I'm doing that. Every week I probably spend a few hours rather than a few days creating nice new words for labels, but then it's the time around the writing that's most productive.

Q **How did you get into the business?**

A I went to university with the three chaps who founded Innocent. I was employee number four. I wasn't writing copy then, more driving vans and delivering smoothies. Gradually I started to write the blurb on the side of the bottles, so I stumbled into it. Before Innocent, I was bumming around Asia as an English teacher. I've always liked writing and thought I was good at it but I didn't know how to make a career out of it. Ending up here and earning my corn as a professional writer was a happy accident.

Q **Innocent is famous for its tone of voice. Did that voice arrive fully formed?**

A Some people think it was a careful marketing decision – it wasn't, it was pretty much unplanned. It grew out of the way we spoke to each other. If it made us laugh then I knew something was halfway to being right. 'Does it make my friends laugh?' was as technical as our market research got.

Q **What was the first thing you wrote at Innocent?**

A It was a label. I came in about a month after the launch and Richard had already written the first eight labels, so I had a go at the next batch.

I think it was one about mangoes – I'd learned something about the particular mangoes we were using, so I did something cliché about being exotic. We can dig it out if you like but I'm sure we'd be underwhelmed.

Q **How did it feel to see that early writing on the shelf?**

A Brilliant. I remember one piece I did about my grandad – of whom I'm very fond – for the back of a tetrapak. I don't think he'd eaten a piece of fruit in his life and we got him drinking smoothies by stealth, so suddenly at the age of 70 he was enjoying fruit. I wrote something about him seeing off the Germans, travelling the world and now getting into fruit. My grandparents stood in the chiller area of their local shop for hours pointing it out to anyone who passed.

Q **Is there a copywriter personality type?**

A I think you need to be a bit of a journalist within our business. There's lots of good stuff happening but it might be buried in a file or a brain somewhere. We have guys travelling the world buying five years' worth of mangoes from India and the like. They've all got great tales to tell but the last thing they want to do when they get back is come over and tell me all about it. So you have to be proactive and put yourself about a bit.

Q **Is there anything fundamentally different about copywriting compared to other forms of writing?**

A Writing here takes a mix of different skills. You've got to go and find the story, so you're a journalist. You've got to make it come alive in people's minds, so you're a poet. So to answer your question, I'm not sure there is. You're still using basic writing skills, but perhaps in a different combination.

Q **Do you ever hanker after doing something – how can I put it – weightier?**

A Well, I'm not a frustrated novelist as some copywriters undoubtedly are. I've met plenty of copywriters and there are always a few budding novelists amongst them. I've got friends who've written novels, and friends who've tried and failed, but I'm less interested in fiction as I get older, I'm more into science and nature.

Q Do you read much?

A I'm married to a woman who reads a lot, so she shames me into reading. I have two young kids so I don't have as much time as I once did. Reading is a luxury now, so I have to know it's going to be good before I start. I mainly read books about the world that will hopefully turn out to be grist to the mill. I'm into space and the cosmos. I have a brilliant Oxford A–Z of food. There are a million things in there I can use in my day job.

Q Any there golden rules for writing at Innocent?

A It's got to be interesting. That's all I do, tell people about interesting stuff. More specifically, here's something I stole about starting to write: you have your first idea – great, but don't use it. Ninety per cent of people will have thought of that, so it'll be boring. Have another idea – great, don't use that either. Eight per cent of people will have thought of that. Go for the third idea – that's what hardly anyone will have got to, that's the one people will say, 'Ah, what an interesting approach'. Plus there's an uber-rule at Innocent to write naturally. It has to sound like a real person speaking.

Q What about your writing routines and rituals? For example, what hours do you work?

A I write at the beginning or the end of the day. I get up early, go for a quick run and get to my desk by 7. Then I can write well until the office gets busy at 9. In the evening after the kids have gone to bed I can do a bit more. I prefer to work in bursts of about an hour. Music is a distraction for me – ideally I need quiet. I'm also a deadline fiend – I need the deadline to be about 3 minutes away before I start, although curiously if it gets too close I go weirdly catatonic.

Q Any writing that you really love or indeed hate?

A I don't massively enjoy writing shelf barkers or anything that veers towards the more commercial. It's got to be done, mind, and some of the best things I've ever written ended up in unglamorous places like the edge of a shelf in Morrisons and sold far more smoothies than some of the wanky ad or digital stuff I've done.

Q How do you plan and prepare?

A We have a wiki here where everyone who learns anything about anything dumps their knowledge. So if I'm writing about blueberries the first thing I do is check the wiki to see, 'Oh, Simon's been in America recently checking the blueberry harvest'. Then I'll go and have a cuppa with Simon and ask him all sorts of stuff. It might turn out he stayed at a weird hotel called the Blueberry Lodge which was shaped like a giant blueberry. That's the sort of thing I'm looking for. There's no point telling people 'we only use the finest blueberries'. They either know that or they don't care. Or both.

Q How long to write a label?

A We get a big bunch of briefs and four weeks later we hand over our stuff, usually a batch of 30 at a time. I can write one in about 15 minutes, and then come back to it a few days later for a bit of spit and polish. I'll usually sit down for an hour and do three or four and give them to Ceri who'll check them and sort them into piles of good and bad, along with the stuff she's written.

Q Do you ever get stuck, and if so how do you break out?

A One thing I do is start from a random word or sentence. So if I've got a brief to write about health and cranberries I might start with a line like 'The thing about squirrels is . . .'. Then you know that however you get back to health and cranberries will be either interesting and funny or completely shit. So it's good to have a few opening lines like that to play with and wake you up.

Q And finally, any words of wisdom for aspiring copywriters?

A Write lots of stuff. Keep writing. Write snappy lines and short stories. Stick your words on a blog. Try writing a bit of everything. But most importantly, just write. It's amazing the number of people who applied for the job of copywriter at Innocent about a year ago who couldn't show me any stuff of note. Also, steal ideas from anywhere – the best CVs that I see are like the start of a good conversation, rather than a list of achievements and medals won for being clever.

In a nutshell:

- A good copywriter is part journalist and part poet.
- Don't use the first thing you think of – everyone else will have thought of it too. Don't even use the second thing you think of. Try using the third instead.
- Be polite – say hello, write to the person you're speaking to, get their attention quickly and leave them with something to think about at the end.
- Don't bother telling people the obvious stuff.
- If you're stuck, set yourself the challenge of getting from a random word back to where you need to be.
- Respect your readers' intelligence.

---

brilliant questions and answers

Jim Davies

'Brilliant copywriting gives the reader a reason to read. It's got to be easy to read and entertaining in a way that effortlessly encourages them to stick with you to the end.'

The taxi driver who picked me up at Leamington Spa station generously provided a running commentary on sites of local interest, including what he proudly claimed was the first polo field to be built in the UK. After I met Jim at his historic pile (complete with priest hole, which I mistook for an airing cupboard) he told me the polo stuff was nonsense and that it had all been built after he arrived in the area a decade ago. Oh well. After lunch Jim ushered me into his office where I couldn't help noticing his D&AD black pencil, recently awarded for work with The Partners on the National Gallery's 'Grand Tour' project. His wife and business partner Deborah brought up tea and we got going.

Q How would you describe yourself?

A I'm a copywriter. I've made a conscious decision to call myself that to differentiate what I do now from my previous job as a journalist. Plus, I'm not so interested in consultancy or training. I really enjoy doing consumer-focused stuff like books, mailers, pieces of print and ads – in fact I get most buzz out of doing short stuff like posters. Rightly or wrongly I describe that as copywriting.

Q How did you get started?

A I did English at university and eventually got taken on at Haymarket Publishing as a sub. At the time they did a design mag called *Direction* where all the cool people worked. I was desperate to get in there, and eventually they hired me as a writer-cum-sub, apparently on the strength of my haircut. Anyway, *Direction* eventually folded but by that time I'd built up lots of contacts in the design industry. Today the only editorial I do is my column in *Design Week*.

Q How useful was that training as a journalist?

A Oh, very. It helped me develop my own style, also subbing is brilliant discipline. In copywriting no one really checks your work for spelling mistakes or typos, so it really helps there. Actually I think the skills are quite transferable and subbing is undoubtedly a good start for a copywriter.

Q You've met lots of writers over the years – what makes a really good one?

A I think they write like they speak, or at least how they'd like to speak. I'm not a particularly good speaker, and if I could change the way I speak it would be to make it far closer to how I write. Good writers are also good observers and they're great at picking things up – the whole magpie thing. They're also adept at going into character, although one of my bugbears at the moment is that everyone is going into the same Innocent-esque character. It's not Innocent's fault but that's what's happened.

Q Any great tips you'd care to share?

A The first sentence is the most important, but the second sentence is the hardest to write. The first line is the hook that has to provoke the reader in some way, but once you've done that you've got to follow it up. That's why the second sentence matters so much, because it's got such a vital supporting role justifying and expanding the first sentence. Also don't leave things to the last minute. I'm one of these people who needs to be at the station 20 minutes before the train leaves.

Q **Any big influences on your writing and thinking?**

A Definitely P.G. Wodehouse. What's great about him is that he's accessible but the sentences are so well constructed and so witty. I'm just constantly amazed by his skill. I'm very impressed by some members of 26, particularly Tim Rich and Will Awdry, and with classic 80s ad people like Tim Delany and David Abbott.

Q **How do you go about planning and preparation?**

A I go over everything they send me, reading and digesting. Then if it's a complex job I get a big pad of paper and jot down ideas in little sentences and phrases. Next I'll highlight any that look promising and I start writing, usually with a headline. I don't order the ideas in advance; instead I might do three of four versions, and show the client the best two – I'm talking about short pieces here, not books obviously. After I've done the headline I'll really concentrate on the second sentence as I said earlier, and build it up from there, subbing as I go and editing very closely. Anything up to around 1000 words I'll write straight out of my head; if it's longer I might make a list of the main points I want to get over and chunk it up.

Q **Let's talk about briefs. The tighter the better?**

A Not necessarily. I like the responsibility of coming up with an idea that's mine rather than one provided by the client. Once you've established trust with a client I quite enjoy finding my own way into a job. I don't like it if they're too specific, because that means there's no room for manoeuvre. The main thing is having enough information. I hate having to call people all the time for information, only for them to tell me, 'Well, you're the writer'. Well, exactly, I'm the writer, not a mindreader.

Q **How do you know when something is working?**

A When Deborah says so. I'm only half joking – it's important to have someone totally honest to review your work. If something isn't quite working it's a good idea to try deleting the first paragraph. Generally you're just warming up and that's just wasting space and scene setting. I tend to get to the meat in the second paragraph.

▶

(Q) And finally, any advice for the aspiring brilliant copywriter?

(A) Just try to inject some personality into it. Inexperienced writers tend to go into this strange formalese when they write copy, so it's about deciding what sort of character this business might be, which in turn is about talking to the client and translating that into natural speech. Sometimes they embrace it, sometimes they don't, but you've got to try. And if you're stuck, try playing The Ramones at top volume. It works for me.

In a nutshell:

- Subbing is superb training for copywriting.
- The best copywriters tend to write like they speak, or rather how they'd *like* to speak.
- Borrow ideas from anywhere and everywhere.
- Learn to mimic, but don't copy the crowd.
- The first sentence is the most important, but the second has to work even harder.
- Go as far as you can creatively, for your own sanity as much as anything. The client will always rein you in if needs be.
- Get someone impartial to review your work.
- Never underestimate the power of a good haircut in securing your first job.

↗ brilliant questions and answers

Tim Riley

'Brilliant copywriting captures the spirit of a brand. A lot of advertising copy feels like it's been written for an awards jury. Brilliant copy isn't just written to entertain you and your mates, it's a commercial thing that aims to sell something to someone.'

Thanks to a two-hour delay caused by badgers undermining a signal box in the Stevenage area I miss my appointment with Tim Riley, Head of Copy at

AMV BBDO London. Graciously Tim makes himself available for a later slot, and we head over the road to a rather splendid old hotel whose cavernous atrium boasts an impressive pot plant jungle and discreetly tinkling cocktail piano. Tea and biscuits arrive, eventually followed by first one then two cups, and suddenly all is well.

Q So Tim, what exactly do you do?

A I'm a writer at Abbott Mead Vickers BBDO. My actual title is Head of Copy – which sounds quite impressive, but in truth it means I get phone calls from account managers asking about the correct use of a semi-colon.

Q Do you do much writing these days?

A I still write every day – print and TV. I don't do much body copy as that's something that only tends to get produced once the client's bought the idea. But yes, I do still write live text, so I'm still a working copywriter.

Q How did you get started as a copywriter?

A I didn't have a clue what to do when I left school so I thought I'd put off making a decision by doing a foundation course. Then I discovered I couldn't draw as well as I thought I could, so I decided to become a graphic designer. Then at art college I discovered I wasn't much good at that either. So I did a D&AD course that involved going to a different agency each week to get a brief, which you'd work on and then get a crit the following week. The best bit about it was that as a writer all you had to do was suggest a headline, while the art director had to do a complete layout. I realised that, for a lazy person, copywriting was the way forward.

Q Was there any senior figure who guided you in that first job?

A To begin with, there was a writer called Alan Curson who would very politely say if something wasn't good enough and advise me to change this or that. Plus my creative director, Alan Tilby, was a copywriter. He'd worked at CDP in the 1960s and was part of the same generation as Charles Saatchi and Alan Parker. People were very generous with their time back then. He used to say we were lucky – when he worked at CDP, *his* creative director – a chap called Colin Millward – had a hole in the wall of his office and you had to hold your ad up and he'd just shout 'yes' or 'no'.

▶

Q **What did you learn at BMP?**

A The basics – keep it short, don't be indulgent. The opposite of writing a novel. With that you take an idea and expand it to 300 pages. For an ad you take maybe five pages of briefing and turn it into one sentence. Plus, I learned that no one is waiting to read what you've written. As somebody writing advertising, you're basically an irritant in people's lives, so you have be as charming as you can.

Q **Just to change tack slightly, is the death of body copy greatly exaggerated?**

A I judged some awards yesterday, and there were lots of ads with plenty of great copy. But would you read those words if you weren't on a copywriting jury? I don't know. I think there's an argument for long copy when there's something vital to say, but those cases are few and far between. In fact the *only* place I really see long copy ads now is when I'm judging awards, so I'd have to say it's in decline. But I don't miss body copy – I didn't like writing it and people didn't much like reading it. Quite a good match really.

Q **Any preference on print vs. digital when it comes to writing?**

A Not really, but I do think that digital hasn't matured yet and no one's quite cracked how to do it. I think that's an amazing opportunity. Think about the first TV ad shown in the UK – a tube of toothpaste encased in a block of ice, on screen for two minutes or whatever, while a voice just droned on. It took another ten years for TV advertising to come of age, and perhaps digital is in a similar position. We haven't really seen its full potential yet.

Q **What constitutes a really good brief for you?**

A I prefer a tight brief but only as long as I agree with it. I was in a meeting once where we were all a bit stuck and someone asked, 'Well, what does the client think the advertising should be?' And the answer was they didn't know, at which everyone sighed and tutted. But that was a good thing. Imagine if they were dictating exactly what we should be doing. I mean, that's why they employed us.

Q Is there anything specific you do to keep fresh?

A You can learn all kinds of things from all kinds of places. There was a piece in *The Economist* – one of our clients – just recently about Barbie's 50th anniversary. Apparently the guy who designed Barbie was a cold war missile designer, which, as the article pointed out, probably explains her breasts. That's not the kind of thing you expect to get from *The Economist*, but there it was. So just be open.

Q And finally, any advice for younger writers?

A One of the most useful people I've ever worked with was John Hegarty, now Sir John. He's not even a writer. But perhaps because he's so good at expressing himself visually, he was very good at getting you to write as concisely as possible. He used to say something like, 'If they could inspire the French Revolution with just three words, we should be able to sell a soap power with less than ten.' It's an exaggeration but it makes a good point.

In a nutshell:

- Copywriting is the opposite of writing a novel. It's about concision, not expansion.
- Be as charming as possible in print.
- Don't dismiss digital – it's still maturing and represents an amazing opportunity.
- Writing is your trade, so learn to just get on with it.
- 'If they could inspire the French Revolution with just three words, we should be able to sell a soap power with less than ten.'
- It's a good thing if the client doesn't know what they want. That's what we're here for.
- Copy other writers' imagination, but never their words.

↗ brilliant questions and answers

Chas Bayfield

'Brilliant copywriting is about what's not said, rather than what is said. It's sweet, it's slick, it makes you happy reading it. It's not stylised advertising copy that reads like it came from an awards annual 20 years ago. Basically it doesn't sound like copywriting.'

I've done all these interviews face to face because that helps the chemistry. But in Chas's case that would have proved rather inconvenient as he's currently 'pottering around' in Tasmania. As well as being the ace copywriter behind the multi award-winning Tango blackcurrant ads and the deliberately cringe-making 'I fancy your mum' campaign for Birdseye Ready Meals, Chas is also the frontman for dirty gospel rocksters The International Christian Playboys, he write songs with Justin 'Hot Leg' Hawkins and he's a founder of Christians in Media.

Q Who are you and what are you?

A I'm a freelance creative – that's what I call myself. The creative side isn't just about copywriting, it's about having ideas. It's about finding the best way to sell a product, and copywriting might be a part of that. It's about using whatever's most effective, and often that involve words on a page, unless I'm doing something purely visual or purely ambient.

Q Let's talk for a moment about that relationship between the visual and the verbal in advertising ...

A The idea that they're in competition is very old school. The days when the copywriter used to craft his words, then slip them under the office door of the art director who was sitting there playing a grand piano, are over. I see it as two people trying to crack the same nut. Ninety per cent of the job is having a great idea; the other 10 per cent is just making sure it happens.

Q So is that traditional relationship on the wane?

A I think the pairing thing is a good idea, it gives you solidarity, it's someone you can be really honest with. I don't think that's going to go away; what I think *will* go away is the very traditional roles of the writer doing the words and the art director doing the visuals. There's so much more that a team can do these days.

Q How did you get started in advertising?

A I was a bit rootless, living in Hamburg and making my living as an English teacher. I taught some account directors in an advertising agency and I got smitten by the environment. I came back to the UK and happened to see a Janet Street-Porter programme called the *Rough Guide to Careers* about life in an advertising agency. Just thought, 'That's me.' So I ended up at Watford College and eventually got taken on at HCL in May 1993.

Q What sort of character type makes a good copywriter?

A The people I respect the most don't seem to have their heads stuck in advertising. They read books, watch TV and live in the real world. Advertising can be a blokey clan where you have to like football, drink beer and wear the right kind of trainers. I don't mean to sound snobbish, but there's a very specific type of person who gets taken on. The people I look up to live in their own crazy world and have a life.

Q I was going to ask you about your top tips, so 'get a life' is clearly one of them. Anything else?

A Go where the people are, eat in McDonald's, read trashy magazines, sit in Starbucks a lot, watch the world, travel on buses, go by coach, not first class, and just realise that the people you need to talk to live in the real world themselves and not in some little media bubble. I can't stand it when people get their ideas from YouTube or style mags or movies. Don't do that – get your inspiration from some weird German novel published 100 years ago, read caravanning magazines or listen to the bizarre things people say on buses.

Q Any particularly big influences on your thinking or work?

A I had a teacher called Roger Pelham when I was 12 or 13 who got me really excited about the English language and what it could do. I remember having to précis 100 words down to 30 words and it was brilliant, like a logic puzzle, and weirdly that's very much what I do now – saying a lot in a few words.

Q **Are you still a big reader?**

A I am. At the moment I'm reading a Czech author who's from the same town my grandparents are from – he's the most famous person it's ever produced. The stuff that really inspires me is beautifully written, particularly Shakespeare and the Bible. The power of Old Testament prophets is awesome – the amount of Tango scripts I wrote based on Haggai and Isaiah 'smiting with their righteous right hand'. No one knew where that came from.

Q **They probably thought it was *Pulp Fiction*, which is your point about ideas being appropriated. Anyway, how do you then focus your thinking down to a concise idea?**

A I've got this thing I call the Magical Ideas Book which I'm constantly topping up with weird stuff I've seen or overheard. It's literally everything I've ever found interesting. The first rule for any copywriter is, what's interesting? If it's interesting people will remember it. Then you need to make sure it's relevant to the brief and ask whether it makes the product the hero. If it does, then it deserves a double tick. So at some point in thinking about the brief what I do is open up the book – which is actually a computer file these days – and think, 'Right, I'm doing something for online bingo, what have I got that's relevant?' And I can pretty much guarantee that I'll get say five ideas that are right on the money. So that's my start.

Q **Do you enjoy writing?**

A I absolutely love it. It's when I'm happiest. I love getting the sentences to balance, which is why I love the Bible and Shakespeare. And *Moby-Dick*, which I think is an amazingly powerful piece of writing. The Guinness surfer ad – which I thought was one of the best-written ads I'd ever heard – got it's dialogue from *Moby-Dick*, so no wonder I liked it, although I secretly wished that they were brilliant at writing instead of being brilliant at borrowing.

Q And finally, any advice for fledgling writers on how to get their career moving?

A Don't obsess about what everyone else is doing. Don't look at what wins awards. Look to life and trust your instincts – try to encapsulate that in all your work. Basically, live among the people you're talking to. And don't get sucked into the blokey, boorish crowd of ad creatives. Just be yourself. Too much advertising is smug without any reason to be. It needs a massive kick up the backside.

In a nutshell:

- In the end, it's all about ideas.
- Use whatever approach will sell the product. Sometimes that involves words on a page, sometimes it doesn't.
- Go where the real people are, do what they do, and look beyond the media bubble for your inspiration.
- Learn from the Bible, Shakespeare and great classic literature – it's the best writing there is.
- Collect all your crazy ideas for use on some future project.
- Believe you're good at what you do.
- Just be yourself and don't obsess about what everyone else is doing.

brilliant questions and answers

Robin Wight

'The way to get me to read 5000 words of copy is to have a headline all about Robin Wight. The more you can make an ad personally relevant to its reader, the more chance you have of getting through. That, for me, is brilliant copywriting.'

Robin Wight – President of The Engine Group and the 'W' in ad agency WCRS – is one of those advertising characters you couldn't make up,

▶

mainly because no one would believe you. Despite being well into his seventh decade he's a true sartorial wonder (recently described as looking like 'Austin Powers' grandad' thanks to his penchant for purple), a serious thinker and a tireless ambassador for advertising. You'll notice that Robin's interview doesn't quite follow the format of the others; instead it veers wonderfully off into brain science, memes and irrationality.

Q **Let me start by asking how you got started as a copywriter ...**

A I sold toothbrushes door-to-door in Liverpool: toothbrushes, pads, scourers and clothes pegs. They told me, 'If you can sell toothbrushes to people who don't clean their teeth, then you can be a copywriter, my son.' A really important part of copywriting is curiosity. I have this phrase, 'Interrogate the product until it confesses to its strengths'; this is my battle cry. So digging into things and using the power of words as a salesman is what attracted me to copywriting. What I've learnt, forty years on, is that a lot of our decisions are made totally irrationally. We have a rational coding mechanism – words – for something that is quite irrational – our decisions. Advertising requires some form of logical argument, but often that argument is used as a rationalisation *after* the buying decision has been made, and copy often just reinforces the belief that you've made the right decision. The biggest readers of car ads are people who've just bought that car. If you understand that we are rationalising machines, not rational machines, then the role of traditional copywriting comes *after* the sale, not before. I think that's an important shift that people haven't really taken on board.

Q **How can a copywriter create that irrational, emotional appeal?**

A At the early stage it's about having an idea that you often express through words, but the notion that someone's brain is divided up into pictures and words is wrong. I think the writing part of copywriting is pretty secondary. When I wrote a piece for *Campaign* about whether the copywriter is dead, I had this headline 'Would David Abbott get a job in advertising today?' And of course he would because he was a brilliant strategic thinker and ideas person – those beautifully caressed words were easier to read than not to read. I think the role of the copywriter today is much more about coming up with the idea, which could even be wordless.

Q So does it make sense to talk about copywriting any more, or are we in the ideas business?

A People's brains have a mixture of capabilities, and what the copywriter ideally brings to the ideas partnership is a brain that is systemised and analytical. Part of the copywriting job is to undertake strategic analysis. In fact what I call the 'strategic planning brain' I think of as part of the copywriting function. Today the actual craft of copywriting is often the smallest part of the job. In television commercials, for example, dialogue is pretty unusual. I do think some of those craft skills are less present than they should be. Too often we rush to the visual solution.

Q Which presumably is why long copy isn't popular in advertising ...

A Yes, you can read a lot of words about a Nike shoe but if you take a photo of that shoe, the visceral image on your brain will make more impact than a logical argument. The right place for copy about that shoe might be in the box. It consolidates your sense of ownership and gives you little soundbites to tell your friends and fuel word-of-mouth. It makes you an ambassador for the brand and gives you raw material for conversation. So the challenge becomes 'How can we create little globules of words that people can spread in a memetic way?' One of the things we're learning from brain science is the power of memes, memory devices that spread words from brain to brain. As copywriters we want to infect people's brains with our brands, and things that have got some memetic quality are more easily caught by the brain than things that just flip through.

Q How can a writer do that?

A You have to believe that your product has something good about it. You might find something that illuminates the whole brand. One thing that we found out recently for BUPA, one of our clients, is that when you go to a BUPA care home they ask you 'What are your dreams?' Which is brilliant – why shouldn't an eighty-year old woman have dreams? When you hear that story, it changes your view of BUPA. Which leads to my next point about copywriting – the power of storytelling. Our brains aren't wired up to receive information as logical arguments; our brains are wired up to hear information as stories. So Kennedy's 'salesmanship in print' is one definition of copywriting, but being able to tell the story of the brand is a more engaging one. 'The future's bright; the future's orange' became a narrative about being future-facing for an entire company. It's a broader approach to copywriting than just salesmanship in print.

Q **Any big influences on your writing/thinking?**

A I've become interested in genetics, and therefore I read about memes in Dawkins' book and heard him talk about memes. As copywriters or whatever we're basically meme men. And when I presented this idea to the agency nobody was terribly interested. Then when we got the 118 118 account I was able to apply meme theory to a particular campaign and it's become incredibly successful. Brain science doesn't make you more creative, but it teaches you how processing occurs and it reminds you how clever the brain is. The brain doesn't want to spend any attention: it's a cognitive miser. That is why long copy is seldom read unless it is very engaging; the brain doesn't have time to process it, it's too busy thinking about love or lunch.

Q **Finally, any advice for someone trying to become a copywriter?**

A Selling things door to door develops your understanding of how people respond. When I was selling door to door it was to help the blind and I had this idea – people said 'You can't do it, it's too sick' – of getting sunglasses and a white stick because I thought my sales would increase. What I'm saying is that you need to think 'How can I get people to engage with this issue and how can I bring it to life?' The other thing I'd say is every day spend 20 minutes investigating something randomly: go to a dictionary, look up a word and find something interesting to say about it. Also make yourself an expert on how the brain works so you know more about the process of communication than the person trying to hire you – without being a smart Alec! Plus your book is all-important, so find people who can help you make your book better: ask them, 'Tell me three things in my book which are good and tell me the things that are bad'. So seek criticism. That way you'll improve.

In a nutshell:

- 'Interrogate the product until it confesses to its strengths.'
- Copy is often used to rationalise a decision we've already made.
- Memes are memory devices that help spread words from brain to brain.
- Our brains aren't wired up to receive information as logical arguments; instead they're wired up to accept information as stories.
- 'As copywriters we're basically meme men.'
- Be brief; the brain is a cognitive miser. It's too busy thinking about love or lunch.
- Sharpen your act by selling stuff door to door.

Bibliography and further reading

The Writer's Voice
Al Alvarez, Bloomsbury, 2005

It's Not How Good You Are, It's How Good You Want to Be
Paul Arden, Phaidon Press, 2003

Whatever You Think, Think the Opposite
Paul Arden, Penguin, 2006

Life's a Pitch
Stephen Bayley and Roger Mavity, Corgi, 2007

The Copywriter's Handbook: A Step-by-Step Guide to Writing Copy That Sells
Robert Bly, Owl Books, 1988

Can I Change Your Mind? The Art and Craft of Persuasive Writing
Lindsay Camp, A & C Black, 2007

The Copy Book: How 32 of the World's Best Advertising Writers Write Their Copy
D&AD, Rotovision, 1994

The Art of Looking Sideways
Alan Fletcher, Phaidon Press, 2001

Get Everything Done and Still Have Time to Play
Mark Forster, Hodder & Stoughton, 2000

How to Get Ideas
Jack Foster, Berrett-Koehler, 1996

Fowler's Modern English Usage
HW Fowler, Oxford University Press, 1990

The Guardian Style Guide
Guardian Books, 2007

The Art of Writing Advertising
Denis Higgins, NTC Business Books, 1965

The Big Idea
Robert Jones, HarperCollins, 2000

Well Written and Red
Alfredo Marcantonio, Dakini Books, 2003

Remember Those Great Volkswagen Ads?
Alfredo Marcantonio, John O'Driscoll and David Abbott, Harriman House, 2008

A Smile in the Mind
Beryl McAlhone and David Stuart, Phaidon Press, 1998

Ogilvy on Advertising
David Ogilvy, Random House, 1998

The Corporate Personality
Wally Olins, Mayflower, 1978

On Brand
Wally Olins, Thames and Hudson, 2004

The Economist Style Guide
Profile Books, 2005

How to Be a Graphic Designer Without Losing Your Soul
Adrian Shaughnessy, Princeton Architectural Press, 2005

Dark Angels: How Writing Releases Creativity at Work
John Simmons, Cyan and Marshall Cavendish, 2006

The Invisible Grail: How Brands Can Use Words to Engage with Audiences
John Simmons, Cyan and Marshall Cavendish, 2006

We, Me, Them and It: How to Write Powerfully for Business
John Simmons, Cyan and Marshall Cavendish, 2006

The Elements of Style
William Strunk, Jr, and EB White, Longman, 4th edition, 1999

Hey Whipple, Squeeze This: A Guide to Creating Great Advertising
Luke Sullivan, John Wiley & Sons, 1998

Rewind: Forty Years of Advertising and Design
Various, Phaidon Press, 2004

InformationAnxiety2
Richard Saul Wurman, Que, 2000

A Technique for Producing Ideas
James Webb Young, NTC Business Books, 1965